George Arthur Merrill

Elementary Text-Book of theoretical Mechanics

Kinematics and Statics. Author's Edition

George Arthur Merrill

Elementary Text-Book of theoretical Mechanics
Kinematics and Statics. Author's Edition

ISBN/EAN: 9783337176969

Printed in Europe, USA, Canada, Australia, Japan

Cover: Foto ©ninafisch / pixelio.de

More available books at **www.hansebooks.com**

ELEMENTARY TEXT BOOK

OF

Theoretical Mechanics

(KINEMATICS AND STATICS)

BY

GEO. A. MERRILL, B. S.

PRINCIPAL OF
THE CALIFORNIA SCHOOL OF MECHANICAL ARTS

[AUTHOR'S EDITION.]

SAN FRANCISCO:
PRESS OF UPTON BROS.
1899

PREFACE.

The subject of Mechanics, as herein treated, is within the comprehension of students in the upper classes of secondary schools. An intelligent journeyman, also, would find no difficulty in reading and understanding nearly every page of this book. While algebra, geometry and trigonometry are freely used to facilitate the demonstration of principles, a person with only a good working knowledge of arithmetic will be able to understand the principles thus deduced and to apply them,—especially by the aid of the graphical methods which are given with some prominence throughout the book.

The author feels that he enjoys almost a monopoly in this field, so far as American publications are concerned. The teaching of Mechanics as a subject *per se* is confined in the main to collegiate courses, and the few American text-books on the subject are written for students familiar with the calculus. A number of more elementary texts have been published by English authors, but their value to American readers is impaired by the fact that they are usually compiled with a view to meeting the conditions imposed by the English examination system, which does not conform to the American educational plan.

The subject-matter has been restricted to Kinematics and Statics. It is left entirely for future consideration as to whether or not a subsequent volume on Kinetics will be issued.

As this is a text-book and not a treatise, it is written from the standpoint of the student, without attempting to force upon him any rigid sequence of topics and ideas that a logical analysis of the subject might seem at times to require. Beyond an effort to abide by a few of the fundamental precepts of teaching, no one method of presentation has been used to the exclusion of others. Only a few experiments are required or suggested. Any good teacher, however, could easily arrange a parallel course of laboratory exercises. In his own classes the author has found that the average student has acquired from his every-day observations and experiences an acquaintance with facts and phenomena quite sufficient to enable him to master the subject without a formal laboratory course. It is probably the same in the higher classes of all schools that include shopwork and general physics in their curricula. Even in the purely academic high schools the course of general physics includes, as a rule, laboratory exercises on the simple machines, friction, acceleration, etc.

The present and prospective prominence of American manufacturing industries not only gives the average young man a zeal for the study of Mechanics, but it also argues the need of giving greater prominence to that subject in the high schools. Looked at from almost any standpoint, the few pages usually alloted to Mechanics in the elementary text-books of Physics are grossly inadequate. Doubtless the teaching of Mechanics as a separate subject would be stimulated by the publication of a number of good text-books in that line.

The author's thanks are extended to Dr. Caroline Baldwin Morrison for numerous suggestions and for the careful reading of the press-proofs.

GEO. A. MERRILL.

SAN FRANCISCO, FEB., 1899.

CONTENTS.

PAGES

INTRODUCTION.
    Fundamental Branches—Kinematics, Statics, Kinetics. Fundamental Ideas—Space, Mass, Time.................................... 1–2

SECTION I—KINEMATICS.

CHAPTER I.   MOTION. VELOCITY.
    Velocity, Uniform Motion, Average Velocity, Relative Motion, Graphical Representation of Motion............................ 3–7

CHAPTER II.   COMPOSITION OF VELOCITIES.
    Parallel Motions, Components at Right Angles, Components at Any Angle, Resolving a Velocity into Components, Triangle of Velocities and Polygon of Velocities, Resultant for Angles Greater than $90°$.............. 8–24

CHAPTER III.   CIRCULAR MOTION.
    Motion of a Body in a Circle, Angular Velocity, Composition of Circular and Rectilinear Motion, Composition of Two Circular Motions, Cycloids........................... 25–37

CHAPTER IV.   ACCELERATION.
    Idea of Acceleration, Definition of Acceleration, Formulæ for Uniformly Accelerated Motion, Acceleration of Gravity.
    Use of Co-ordinate Axes, Graphical Representation of Distance Traveled by an Accelerated Body.
    Projectiles, Elevation and Range, Maximum Range................... 38–56

SECTION II—STATICS.

		PAGES
CHAPTER I.	FORCE. MASS.	
	Different Kinds of Force, Definition of a Force, Resistance, Statical Conditions, Action and Reaction, Composition and Resolution of Forces, Law of Gravitation, Mathematical Expression of the Law of Gravitation, Weight of a Body Beneath the Earth's Surface, Mass, Density, Heavy and Light Bodies Fall at the Same Rate..........	57-73
CHAPTER II.	WORK. POWER. ENERGY.	
	Foot-Pound, Horse-Power, the Watt, Energy, Potential Energy, Kinetic Energy, Molecular Energy and Mechanical Energy, Chemical Energy, Transference and Transformations of Energy, Conservation of Energy, Graphical Representation of Work, Indicator Diagram.	74-93
CHAPTER III.	CENTER OF GRAVITY.	
	Center of Figure, Center of Mass, Center of Gravity, To Locate the Center of Gravity of a Body, Equilibrium, Stability of Equilibrium, Degree of Stability............	94-108
CHAPTER IV.	PRINCIPLES OF MACHINES. THE LEVER.	
	Tools and Machines, Efficiency of Machines, the Simple Machines or Mechanical Powers, the Lever, Principle of Virtual Work, Principle of Moments, Moment Due to Weight of Lever, Three Kinds of Levers, Compound Levers, Safety Valves, Pressure on the Fulcrum, Parallel Forces, the Couple, Bent Levers, Moment of a Force Acting Obliquely on a Lever, Mechanical Advantage..........	109-131

CONTENTS.

CHAPTER V. MACHINES.
Wheel-and-Axle, Gearing and Shafting, the Pulley, Fixed Pulley, Movable Pulley, Combinations of Pulleys, the Inclined Plane, the Wedge, the Screw, Endless Screw, the Cam and Eccentric, the Toggle-Joint, Differential Motion, Compound Machines........ 132–164

CHAPTER VI. FRICTION.
Sliding Friction, Coefficient of Friction, Static Friction and Kinetic Friction, Friction Always a Resistance, Determination of the Coefficient of Friction by Means of the Limiting Angle or Angle of Repose, Work Done in Dragging a Body by Sliding, Rolling Friction, Work Done in Dragging Vehicles, Anti-Friction Wheels and Ball Bearings, Lubricated Surfaces, Friction of Ropes Belts and Cables, Measurement of Power Transmitted by Belts, Methods of Increasing the Efficiency of Belts 165–195

THEORETICAL MECHANICS.

INTRODUCTION.

The scope of Theoretical Mechanics is indicated in the following outline of its **fundamental branches**:

I. The branch called **Kinematics** is a study of the motions of bodies without reference to the cause of their motion; it inquires *how*, or in what manner, a body moves, as distinguished from *why* it moves.

II. The branch called **Statics** is a study of pressures, etc., in bodies;—that is, the influence of forces that are unable to move the body acted upon, because of other balancing forces or resistances. When an internal resistance balances an exterior force the body is said to be under "stress."

Pressures in fluids give rise to special considerations, which are grouped under a subdivision of Statics called **Hydrostatics**.

In the study of Statics, geometric and graphic methods are applied so widely, as in computations pertaining to roof trusses, bridges, etc., that it has been found convenient to give this mode of treating the subject a special name, **Graphostatics**.

III. The third branch deals with the relation between the motion of a body and the force or forces producing the motion, and is called **Kinetics**. It considers *why* the body moves, and the relation between the manner in which it moves and the influences causing the motion.

It will be observed that in Statics and Kinetics we have to deal with forces. When classed together on that basis these two branches are sometimes called **Dynamics**, in contrast with Kinematics, or the study of pure motion.

Fundamental Ideas. The principles of mechanics are deduced and developed by mathematical operations, based upon the fundamental ideas of space, time and mass. The mathematical operations used in an elementary treatment of the subject are mainly arithmetical, but are greatly facilitated by the use of Algebra, Geometry, and Trigonometry. The ideas of space, time, and mass are usually presented in the earlier parts of an elementary course of physics.

Measurements of length, area, volume and capacity, and also angular measurements, are all comprehended in the idea of **Space**.

Although a rigid definition of **Mass** is rarely required outside the subject of mechanics, the related ideas of weight, density, and specific gravity are frequently met with.

The unit of **Time**, a second, is well understood as meaning a certain fraction of a day, but it should be remembered that a "day" is not always of fixed length. If the earth had but one motion, a rotation upon its axis, the successive transits of the sun across any meridian (noonday) would always occur at equal intervals. But since the earth also revolves around the sun, and in an elliptical rather than a circular orbit, the interval between transits is not constant. For scientific accuracy this difference must sometimes be allowed for; but for most purposes it is sufficient to take as the length of a "mean solar day" the average of these successive intervals of transit of the sun across any meridian. The ordinary second is $\frac{1}{60 \times 60 \times 24}$ of this mean solar day.

Section I.

KINEMATICS.

CHAPTER I.

MOTION. VELOCITY.

Velocity. When a body starts to move from a given point a full description of its motion involves:

1. Its direction of motion;
2. Its rate of motion.

The idea of direction is derived from the conception of space, and the "rate of motion" implies, in turn, the other two fundamental ideas,—distance and time. By the rate of motion we mean the distance traveled in a unit of time. This is usually called the **Velocity** of the body.*

Examples:

1. (*a*) *A velocity of 10 meters per sec. is equivalent to how many feet per second?*

(*b*) *Express the same velocity in feet per minute.*

2. (*a*) *Express in feet per sec. a velocity of 100 meters per minute.*

(*b*) *Express the same velocity in feet per hour.*

---

* The **velocity** of a body, when considered independently of the direction of motion, is sometimes called the **speed** of the body.

3. (a) *A velocity of 1000 meters per min. is equivalent to how many miles per hour?*

(b) *Express the same velocity in feet per sec.?*

4. *Which velocity is the greater and by how much,—40 miles per hour, or 12 meters per sec.?*

Uniform Motion; Average Velocity. The distance traveled in any given time by a body moving with uniform velocity is expressed by the formula

$$d = v\,t. \tag{1}$$

By transformation this formula becomes

$$v = \frac{d}{t}, \tag{2}$$

which indicates that the velocity of any moving body may be found by dividing the distance traveled by the units of time consumed. (Compare this statement with the definition of velocity, already given.) This is true even if the speed of the body is not uniform, the quotient d/t representing in this case the average velocity during the time under consideration.

Example:

A train travels 10 miles at a rate of 20 miles per hour; then 4 miles at an average rate of 30 miles per hour; then 6 miles at a uniform rate of 40 miles per hour; then it comes to the rest after traveling a distance of 1 mile (slowing down), running meanwhile at an average rate of 20 miles per hour; it stands for 7 minutes; then it starts and runs for 20 minutes at the average speed of 21 miles per hour. What has been its average velocity for the entire time?

Relative Motion. An object at rest on the earth's surface may or may not be in a condition of absolute rest. It moves with the earth in diurnal rotation, and it progresses with the earth in its path of revolution around the sun, but there may be other motions due to the sweeping of the solar system through space, the effect of which may be to bring the object under consideration to a condition of absolute rest for the instant. More probably, these unknown influences add to the complexity of the actual motion, but *relatively to the earth* the body continues at rest.

Examples:

1. *Two trains pass each other at a station, called A. One train, called E, is moving eastward at the rate of 30 miles per hour; the other, W, is moving westward 25 miles per hour.*

Find:—(a) The velocity of E relative to A.
    (b)   "   "   " W "   " A.
    (c)   "   "   " E "   " W.
    (d)   "   "   " W "   " E.

DISCUSSION:—In this example, if we consider only the *rate* of motion, then, since the train E would be located 30 miles from the station at the end of an hour, it would be sufficient to say that its velocity relatively to the station is 30 miles per hour. But, if we also consider *direction* of motion, then we must add that the motion of the train E relatively to the station is in an Eastward direction.

Likewise the motion of W relatively to A is 25 miles per hour Westward.

At the end of the hour the train E is 55 miles East of the train W. Without giving to the train W any credit for its part of the transaction, we simply recognize the fact that in the hour of time E reaches a position of 55 miles East of W, and we express this by saying that the velocity of E relatively to W has been 55 miles per hour Eastward.

Conversely, we must say of W, that its velocity relatively to E has been 55 miles per hour Westward.

6   THEORETICAL MECHANICS.

2. *A train 400 feet long travels between two mile-posts at a uniform rate of 30 miles per hour.*

At position 1 *a person starts from the last car to walk through the train, and reaches the front at the instant the train reaches* 2.

(*a*) With what average velocity in miles per hour did he walk?

(*b*) What was his average velocity relatively to the ground?

(*c*) What would have been the answers to (*a*) and (*b*) if he had walked to the back from the front of the train?

Graphical Representation of Motion. The motion of a body at any instant,—both the magnitude and the direction of its velocity,—can be represented by a straight line. For this purpose we can adopt any convenient scale of magnitude and any arbitrary notation to indicate direction. For instance, a velocity of 20 miles an hour in an eastward direction could be represented by a line 2 inches long, (1 inch = 10 miles per hour), drawn horizontally from left to right from the starting point, the direction being indicated by an arrow, as OA in the following figure:

If in connection with this motion we wish to refer to a second motion,—say a velocity of 15 miles an hour northward from the same point,—we would have to adhere to the same scale, (1 inch = 10 miles per hour), and instinctively we would draw this line vertically upward.

Exercises:

1. *Using the same notation represent each of the following motions, using the same starting point for all:*

22 miles per hour Southward.
5   "   "   "   Westward.
17   "   "   "   N. E.
14.5 "   "   "   S. E.
3   "   "   "   65° S. of E.
20.5 "   "   "   77° N. of E.
11.7 "   "   "   14° W. of N.
11.7 "   "   "   33° W. of S.

It should be carefully noted that the scale of magnitude in this figure is not a scale of distance, and the lines themselves do *not* represent *distances* merely, *but velocities.* A similar method could be, and frequently is, used for distances, and hence it becomes necessary to keep clearly in mind the real signification of each diagram, and each line in it. Take for example, the confusion that might arise in interpreting a diagram constructed as follows:

2. *On a scale of 1 inch = 10 meters, construct a diagram to represent the* PATH *of a body moving as follows: From the starting point the body moves westward 10 seconds, at the rate of 3 meters per second; thence northward 2.4 seconds at the rate of 5 meters per second; thence north-east 7 seconds, at the rate of 1.7 meters per second; thence 30° east of south for 6 seconds at the rate of 2 meters per second.*

Having constructed this diagram we can use it for any measurement of *distances* concerned with the path of this body, but if we wish to consider incidental questions of velocity it becomes necessary to introduce additional computations involving the consideration of *time.* For instance, at the end of the motion how far was this body from the starting point? And if a second body had moved in a straight line from the starting point to finish, instead of by the roundabout path, at what rate should it have traveled in order to arrive at the terminal point simultaneously with the first body?

It is true that a line which represents a velocity may be regarded at the same time as representing distance, provided we remember that it is *distance per unit of time.*

CHAPTER II.

COMPOSITION OF VELOCITIES.

Sometimes, and generally, the actual motion of a body is the result of several different influences acting simultaneously, as in rowing a boat in a tidal current. Heretofore, in speaking of the velocity of a body we have thought of it only as a simple motion in one direction, taking it as we find it, no matter how many different influences may have contributed to give the body this velocity and path. When two or more influences act *simultaneously* to produce motion in a body each influence has its full effect in its own direction, as if the other influences did not exist. The motion that each of these influences would produce is called a **component**, and the actual motion, as we have been considering it, is called the **resultant** of all these components. There are times when the component motions are known and the resultant cannot be found except by computing it from the components; at other times the actual motion is given and it is desired to find its component parts. We have only to take the literal meaning of "component" to see that the resultant is obtained by "putting together," or combining, these more elementary motions, but the mathematical process of doing this is not always simple, although sometimes it is a mere addition or subtraction.

For the convenience of the beginner it is well to assume three cases, exemplified as follows:

(i) If a person walks directly forward through a moving car his resultant is his rate of walking *plus* the velocity of the train. This would be the composition of motions parallel to each other.

COMPOSITION OF VELOCITIES.

(ii) If he walks straight across the car his resultant velocity is the hypothenuse of a right triangle, of which one side represents the rate at which he walks and the other side represents *on the same scale* the velocity of the train. This is illustrated in Fig. 3. The man starts from A to move across the car toward the window W. Suppose that in a unit time he covers that part of the distance represented by AB: then the arrow AB represents his velocity of walking.

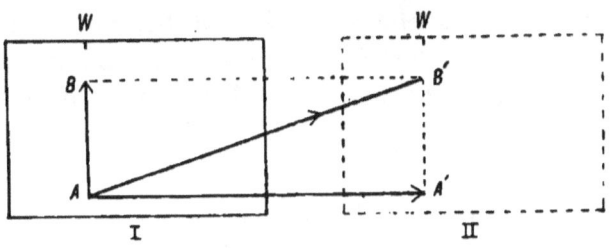

Fig. 3.

Now, since the car is in motion, suppose that it progressed from position I to position II while the man walked from A to B,—that is, in a unit of time. Then the arrow BB' will represent the velocity of the car,—or any point in it, as B. Hence, at the end of the unit of time the man finds himself, not at B, (as he would if the car were at rest), but at B', and the actual path he has traveled in space,—or, better, relatively to the ground,—is AB'. This hypothenuse, therefore, represents his resultant velocity, on the same scale as AB.

Properly speaking, the two components of AB' are the two motions of the man himself,—one AB, relatively to the car, (which, also, would have been his only motion relatively to the ground if the train had not been moving when he walked across); and the other AA', which would have been his only motion relatively to the ground had he stayed in his seat while the train moved. These two motions occurring conjointly, his resultant has a direction between the two. In finding the magnitude and direction of AB' it makes no difference whether we use the triangle ABB', or its equal $AA'B'$; each is half the rectangle $ABB'A'$.

(iii) If he walks obliquely across the car towards a point X, as in Fig. 4, his resultant motion relatively to the ground will be the side AB' of the obtuse triangle ABB'. This case differs from the last only in the obliqueness of the two components AB and AA', but this difference

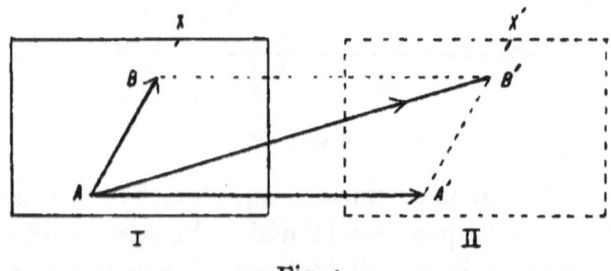

Fig. 4.

is a very important one from a mathematical point of view. It is only when the angle BAA' is 90°, or 45°, or 60°, or 120°, that the triangle ABB', or $AA'B$, can be readily solved without using trigonometry.

Composition of two motions parallel to each other. In Example 2, page 6, as the man walks through the train, he is subject to the additional motion of the train itself. In part (*a*) of the problem he covers in two minutes a distance of 400 feet (by walking) +5280 feet (by the progress of the train). These two component motions take place simultaneously, and since they are in the same direction the resultant velocity of the man relatively to the ground is the sum of the two components; $v = d/t = \frac{5280+400}{2}$ feet per minute.

Observe that the two motions occur in the same interval of time; if they had been *consecutive*, instead of *concurrent*, the conditions would have been very different and would not have come within the scope of "composition of velocities." That is, if the man had remained in his seat at the rear of the train until the train came to rest at mile-post [2], he would have covered a distance of only 5280 feet in the two minutes; if, thereafter, he had walked to the front of the train while it was still in waiting at mile-post [2], he would have required two minutes additional time in which to walk the length of the train, 400 feet. His velocity during the first two minutes would have been $\frac{5280}{2}$ feet per minute, and during the second two minutes, $\frac{400}{2}$ feet per minute. If there were any need to know his *average velocity* during the total of these two intervals of time it would be $\frac{5280+400}{4}$ feet per minute, but his motion during the second two minutes has no effect to change the rate at which he covered ground during the first two minutes, and *vice versa*. In other words the "composition of velocities" is concerned only with the resultant of two or more components, acting conjointly or simultaneously.

The resultant of two or more **parallel motions** in the same direction is equal to the sum of the components; if the two motions are in opposite directions their resultant is equal to their difference. Using the signs + and − to indicate opposite directions, we can say, in general terms, that the resultant of two parallel motions is equal to their algebraic sum.

A person rowing directly with or against wind and tide is subject to three influences, each of which if undisturbed, would cause him to move with a certain velocity. His resultant or actual motion could be found by summing up the three motions that would have been occasioned by these influences acting singly.

Examples:

1. *In Example 2, page 5, if the man had walked to the back of the train from the front, as in part "c," what would have been his resultant velocity?*

2. *In Example 2, page 7, were the motions designated as simultaneous, or as consecutive?*

Components at Right Angles.

The statements in (ii) page 9, will aid in the solution of the following

Examples:

1. *If a man undertakes to row straight across a channel in which there is a current, his course will be oblique.*

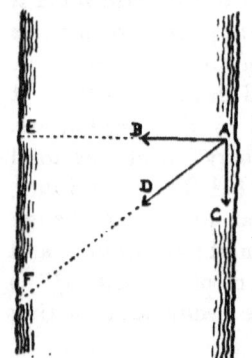

Fig. 5.

Let AB = his velocity of rowing, and AC = his velocity with the current.

Then the resultant AD, the diagonal of the parallelogram, will be the direction of his course.

(a) If $AB = 4$ miles per hour, and $AC = 3$ miles per hour, what is his resultant velocity?

(b) If the DISTANCE straight across the channel (AE) is 8 miles, what is the distance AF?

How long does it take him to reach F?

How long would it have taken him to reach E, if there had been no current?

(c) Find the value of the angle DAB, by trigonometry.

2. *If he wants to go straight across he must head the boat up stream, as AB'.*

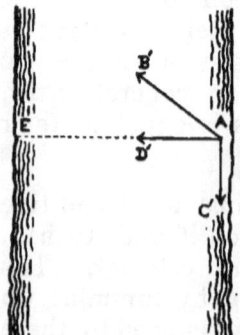

Fig. 6.

Let $AC' = 3$ miles per hour, as before, and let the DISTANCE $AE = 8$ miles.

In what direction and with what velocity must he row in order to reach E, directly opposite A, in 2 hours?

3. *A boat B, 300 yards from shore and 100 yards up stream from A, is carried down stream 3 miles per hour.*

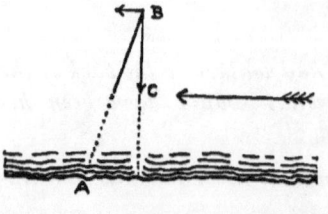

Fig. 7.

(a) If it is kept headed perpendicular to the shore with what velocity must it be rowed in order to land at A?

(b) What will be the resultant velocity?

(c) What is the magnitude of the angle ABC?

COMPOSITION OF VELOCITIES. 13

4. *A steamship, S, 6 miles from the shore, AB, is supposed by those on board to be sailing from south to north 10 miles per hour, but really drifts towards shore at the rate of 2 miles per hour. The shore extends north and south.*

(a) How long before the steamship will reach the shore? Designate the point at which it will strike. With what velocity will it strike?

(b) If it is desired that the steamship shall progress exactly northward at the rate of 10 miles per hour, in what direction and with what velocity must it steam in order that the action of its engines combined with the influence of the current may give it the desired resultant motion?

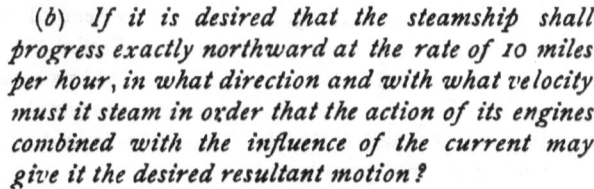

Fig. 8.

Note how these examples illustrate the statement, previously made, that when two or more motions act simultaneously to produce motion in a body, each influence has its full effect in its own direction, as if the other influences did not exist. In Example 4(*a*) it will be noticed that the ship drifts towards the shore at a certain fixed rate, no matter how fast it is sailing in a direction parallel to the shore; and conversely, its progress from south to north is not influenced by the drifting shoreward. Likewise, in Example 1 the current does not retard the man's progress across the channel, but simply carries him down stream.

Two Components at Any Angle. This is the general case and requires the use of trigonometry for solution, although results sufficiently reliable for most purposes can be obtained by graphical methods from carefully constructed diagrams.

Suppose, in a given case, that a body is subject to two simultaneous motions, one 120 feet per second and the other 70 feet per second, the directions of the two differing by 60°. For graphical representation assume a scale of 1 inch = 100 feet per second. Let O be the starting point of the body. Draw OA equal to 1.2 inches, to represent one of the velocities, and OB, equal to 0.7 inch, to represent the other, the angle or difference of direction between them being made equal to 60°. From point A as a center describe an arc whose radius is equal to OB, and from B as a center with radius equal to OA describe a second arc intersecting the first at C. The figure $OACB$ is thus a parallelogram. Its diagonal OC represents the resultant of OA and OB in both magnitude and direction, on a scale of 1 inch = 100 feet per second.

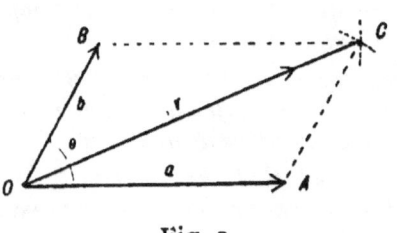

Fig. 9.

Calling these components a and b respectively and the resultant r, it can be shown by trigonometry that

$$r = \sqrt{a^2 + b^2 + 2ab \cos \theta}, \qquad (3)*$$

where θ is the angle between a and b.

It can also be shown that this formula is the same whether θ be acute or obtuse.

Examples:

1. *A boat sailing westward 13 miles per hour is carried by the tide in a direction 23° W. of S. at the rate of 3.5 miles per hour. Find its resultant motion (velocity and direction).*

2. *A bird flying in a N. E. direction at the rate of 28 miles per hour encounters a wind blowing from a direction 12° W. of N. which carries him out of his course at the rate of 19 miles per hour. Find his resultant velocity and direction.*

3. *In the formula* $r = \sqrt{a^2 + b^2 + 2ab \cos \theta}$, *if* $\theta = 180°$ *what is the value of* r? *What is the value of* r *when* $\theta = 0°$? *When,* $\theta = 90°$?

---

* It should be carefully noted why this formula differs from the formula for the trigonometric solution of a triangle, given two sides and their included angle.

COMPOSITION OF VELOCITIES.   15

Resolving a Velocity into Components.

Example:

If a body is moving in a direction exactly N. W., 4.5 miles per hour, how fast is it progressing northward, and how fast westward? Solve this by constructing a diagram to scale, and also by trigonometric computation.

In this instance we have assumed the given velocity to be a resultant, and have resolved it into two components, one north and the other west. We might have resolved it into other components,—two components in directions other than north and west, or even into several components.

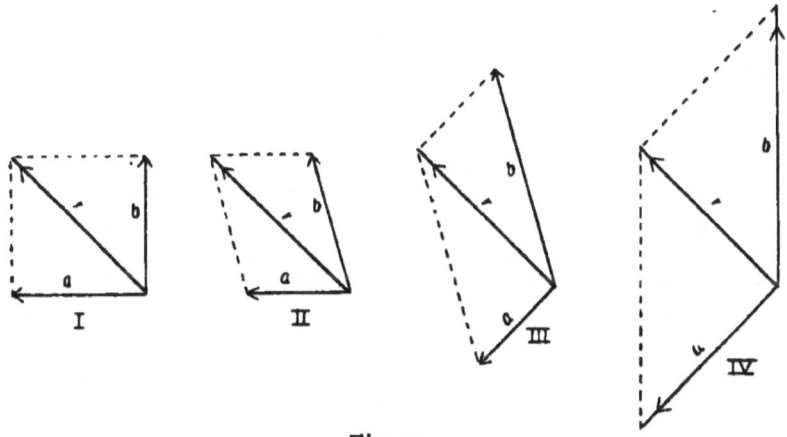

Fig. 10.

Starting with the given velocity of 4.5 miles per hour in a N. W. direction, we can draw an infinite number of parallelograms of which this line will be a diagonal. In Fig. 10, the two sides, *a* and *b*, of any one of the parallelograms may be taken as the two components, of which the actual motion of 4.5 miles an hour in a N. W. direction is the resultant. Sometimes, as in the case of a ship sailing in an invisible current, it is utterly impossible to discover the true components which have operated to give the resultant motion, but as a rule sufficient conditions are known from which to determine what the elementary motions would have been of each if the component influences had acted singly.

To resolve a resultant into two components requires the solution of a triangle, (as in the converse proposition of finding the resultant from the components), and hence at least *three* parts of the triangle must be known. If we start with the resultant given, the conditions of the problem must state, or imply, (1) the lengths of the other two sides; or (2), one of these sides and an angle; or (3) two of the angles.

The resolution of a velocity into components was illustrated in Examples 2 and 3, and the last part of Example 4, pp. 12 and 13.

This triangle to be solved always contains *one of the components* and *a side parallel and equal to the other component*, but it never contains both components. This fact frequently occasions confusion in the graphical representations that accompany this branch of kinematics, and leads beginners into many difficulties in gaining a clear conception of the idea of the relation between a resultant and its components. Remember that the resultant is always the diagonal of a parallelogram of which the components are two sides; and that the two components and the resultant all start from the same point.

Exercises:

Solve by trigonometry. Also, as a check upon errors, construct diagram on convenient scale in each case. Those who have not studied trigonometry can get sufficiently reliable results from diagrams.

1. *A ship is sailing at the rate of 10 miles per hour and a sailor climbs the mast 200 feet high in 30 seconds. Find his velocity relative to the earth.*

2. *A balloon leaving the ground ascends with a vertical velocity of 70 feet per second and is carried with the wind. If it rises at an angle of 83° with the horizon, at what rate is it moved by the wind?*

3. *One of the rectangular components of a velocity of 60 miles per hour is a velocity of 30 miles per hour; find the other component.*

4. *The components in two directions of a velocity of 30 miles per hour are velocities of 15 and 25 miles per hour; determine their directions.*

5. *A steamship is headed in a direction 38° S. of E. in a wind blowing from a direction 27° 30' E. of N. If the ship's progress proves to be in a direction 43° S. of E. with a velocity of 17 miles per hour, at what rate was it steaming, and at what rate was it carried with the wind?*

6. *Find the horizontal and vertical components of the following velocities:*

(*1*) *1000 feet per second in a direction inclined 30° to the horizon.*

(*2*) *The same velocity in a direction inclined 50° to the horizon.*

(*3*) *25 miles per hour at 60° to the vertical.*

7. *Find the magnitude and direction of the resultant of the following motions (four components), to which a body is subjected simultaneously:*

10 feet per second, E; 7 feet per second, N; 13 feet per second, W; and 16 feet per second, S.

8. *Solve by Geometry:*

(*a*) *Given two equal components at an angle of $\beta = 60°$. Find the resultant.* (Hint: The diagonals of a rhombus bisect each other at right angles.)

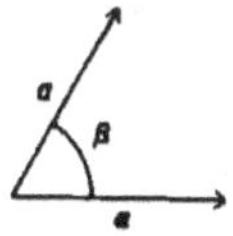

Fig. 14

(*b*) *Find the resultant of two unequal components, a and b, at an angle of 60°.*

(*c*) *Find the resultant of two equal components at an angle of 120°.*

(*d*) *Find the resultant of two unequal components at an angle of 120°.*

9. *In formula 3, p. 14, substitute for θ the above values, 60° and 120°, and compare with the geometric results.*

18   THEORETICAL MECHANICS.

10. The Ship Problem. *A ship is pointed W. to E., with her mainsail set at an angle of 20° (δ in diagram), Fig. 11-a. The wind blows at an angle of 75° with the sail (indicated by β in diagram), with a velocity of 39 miles per hour.*

(a) What is the component of wind velocity in the direction of the ship?

DISCUSSION:—This question is not intended to imply that the motion of a ship is caused to any great extent by the component of the wind velocity parallel to the direction in which the ship is headed. The main motive power comes from the action of the wind on the sail; if we consider only the wind components parallel and perpendicular to the hull we are dealing merely with minor influences,—one acting on the stern of the hull pushing the ship forward, and the other acting at right angles to the hull and pushing the ship out of her course. Although these influences have little to do with propelling the ship, still it is well to study them and put them aside before considering the action of the wind on the sails.

Notice, also, that the question asks only for the wind component in the direction of the ship. This cannot be found from the conditions of the problem unless we assume something concerning the other component. The solution of a triangle requires that at least three parts be given. In this case if OA, Fig. 11-b, represents the

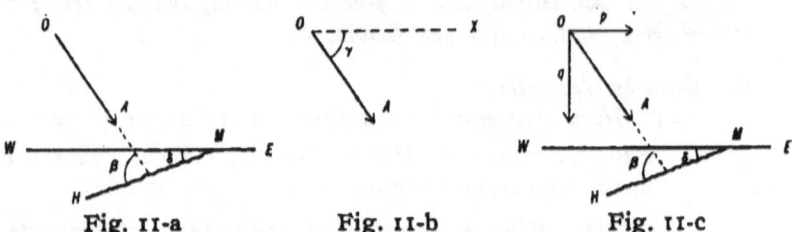

Fig. 11-a   Fig. 11-b   Fig. 11-c

wind velocity in magnitude and direction, and a horizontal line, OX, is parallel to the direction of the ship, we have as given conditions only the line OA and the angle γ. The magnitude of the horizontal component may be almost anything,—depending upon what conditions are attached to the other component. (See pp. 15 and 16).

When not otherwise specified it is customary to take the two components at right angles to each other. In this case, when asked to find the component parallel to the direction of the ship, we have assumed that the other component is at right angle to the ship. (Fig. 11-c).

(b) *What are the wind components parallel and perpendicular to the sail?* (*Fig. 12-a*)

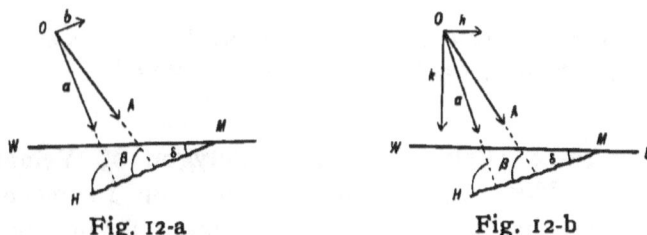

Fig. 12-a   Fig. 12-b

(c) *Now the component* a, *pushing square against the sail, is not entirely useful in moving the ship forward; on the contrary the ship tends to move bodily in the direction of component* a. *How much is component* a *acting in the direction of the ship, and how much is it acting perpendicular to that direction?* (*See Fig. 12-b*).

11. The Kite Problem. *A kite flying in a wind blowing 22 miles per hour, is inclined at an angle of 58° with the horizon, as angle β in Fig. 13-a.*

(a) *Assuming that the wind is blowing horizontally, what are its components—*a *perpendicular, and* b *parallel, to the kite?*

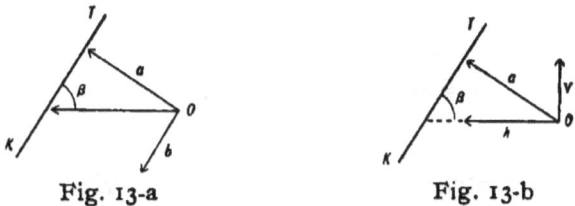

Fig. 13-a   Fig. 13-b

(b) *The component perpendicular to the kite represents the total pull against the string exerted by the wind, (assuming that the string is perpendicular to the kite). A part of this pull is upward and a part horizontal. Find each part,* h *and* v, *Fig. 13-b.*

This part *v*, determined in this manner, is the lifting component of the wind on the kite.

12. *If a body travels* d *miles in* t *hours, what is its velocity in feet per second?*

13. *If a body has a velocity of* v *feet per second, in how many hours will it travel a distance of* m *miles?*

14. *Find the resultant of two perpendicular components, one* p *meters per minute, and the other* q *feet per second.*

"Triangle of Velocities" and "Polygon of Velocities." These two ideas find frequent expression in mechanics. The former has been referred to incidentally in a previous paragraph, and is now reverted to for the purpose of directing attention to a consideration that will serve to facilitate the extension of the subject of composition of velocities to cases in which there are more than two components.

As already stated, the resultant of two components, a and b, is one of the diagonals r of a parallelogram $OACB$ (Fig. 15). In problems concerned with a, b and r, it is

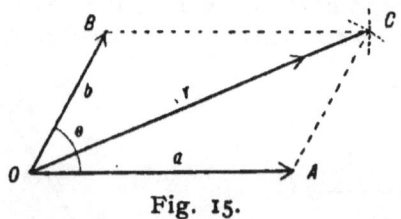

Fig. 15.

customary to solve one of the triangles, OBC or OAC. If we work from the triangle OBC we do not deal directly with component a, but with BC, parallel and equal to a; or, if we choose to work from triangle OAC, we consider AC in place of component b.

Now, what is signified by this substitution of a line parallel and equal to one of the components for the component itself?

COMPOSITION OF VELOCITIES.

It has already been emphasized (p. 11) that the very idea of the composition of velocities presupposes that the component actions take place simultaneously. In the figure of the parallelogram this conception of simultaneous action is easily comprehended; but it is abandoned for the time being as soon as we transfer our thoughts to one of the triangles alone. For instance, take the triangle OAC apart from the rest of the parallelogram. The resultant is found

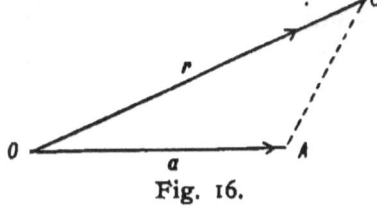

Fig. 16.

mathematically from OA and AC, Fig. 16, as if resulting from these two motions taken consecutively, whereas, in truth, the component motions are not only *not consecutive*, but AC is not one of them at all. This misconception grows out of the tendency, already referred to, to assume that OA and AC represent *distances* or *displacements*, instead of *velocities*. The "triangle of velocities" should be used with a clear understanding of the meaning of each line.

As a mathematical expedient, the "triangle of velocities" is a simple case of a more general construction, the **polygon of velocities**, which suffices not merely for two, but for any number of components. The idea is simple. Let p and q (Fig. 17-a) be two components with resultant r.

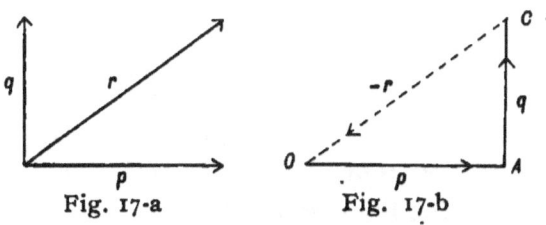

Fig. 17-a.   Fig. 17-b.

Now, if we draw a line OA (Fig. 17-b) to represent p; from A, a second line AC parallel and equal to q; and thence from C a line CO to the starting point, we will have constructed a closed figure of which the closing line is *parallel and equal to* the resultant *r*, but drawn in the *opposite direction*. For the sake of convenience, we temporarily lay aside all thought of simultaneous action and consider the components successively, and we thereby reach a conclusion which we know bears a certain relation to the resultant,—a line equal to the resultant but drawn in an opposite direction.

A similar device, really the same process, can be used for any number of components. Let *p*, *q*, *s* and *t* (Fig. 18) be four components. By resorting to the parallelogram

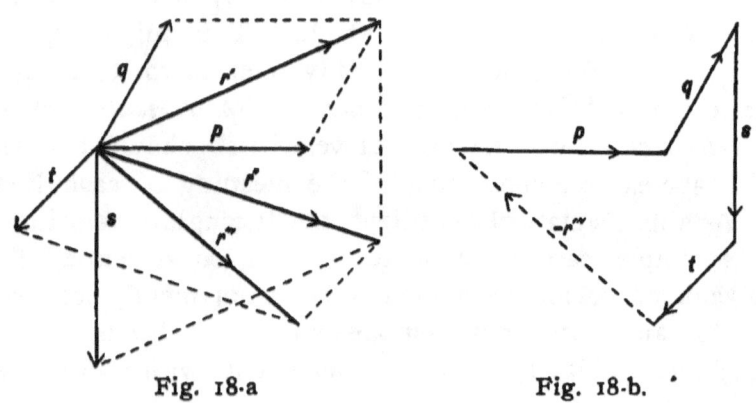

Fig. 18·a.   Fig. 18·b.

method we can get the resultant *r'* of *p* and *q;* then we can treat *r'* as a single motion and combine it with *s* in the same way, getting a new resultant *r''*; this in turn can be combined with *t* to obtain the final resultant *r'''*.

By using the **polygon of velocities** we could have determined this resultant much more readily. Draw a line equal and parallel to *p* (Fig. 18-b); from the extremity of

COMPOSITION OF VELOCITIES.

this draw a second line equal and parallel to q; thence a line equal and parallel to s; and one equal and parallel to t. The line necessary to close this polygon will be *equal to the desired resultant but opposite in direction.*

While the "polygon of velocities" simplifies the graphical determination, it would still be necessary to break up the figure into triangles in order to accomplish a trigonometric solution.

Examples:

1. *How would it have been if we had combined the components of Fig. 18-a in a different order? Try it, assuming any four velocities and finding their resultant by combining them graphically in at least two different orders by the parallelogram method.*

2. *If the lines of Fig. 18-b were constructed in a different order, would the result be the same?*

3. *Find the resultant motion of a body which has the following component motions:*

   8 miles per hour from W. to E.;
   3 miles per hour $34°$ W. of S.;
   5.5 miles per hour $11°$ S. of W.

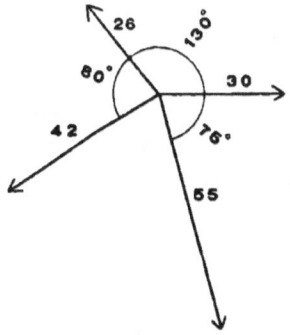

Fig. 19.

Construct the "polygon of velocities" to accurate scale, and compare the graphical result with the trigonometric solution.

4. *A body has four motions measured in meters per second, as indicated by the numbers in Fig. 19.*

Find the resultant by construction and by computation.

Resultant for Angles Greater than 90°. It is frequently asked, "How can the resultant be less than one of the components?" This is often the case when the angle between two components is greater than 90°. Let the two components be *a* and *b* (Fig. 20-a). Now, for convenience,

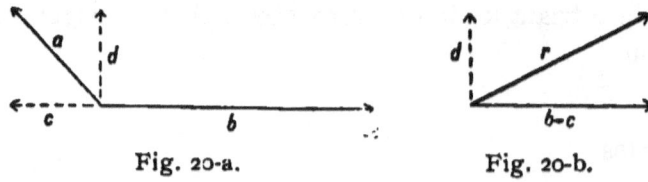

Fig. 20-a.　　　　　Fig. 20-b.

we may project *a* horizontally and vertically,—that is, regard it as the resultant of two components *c* and *d*, at right angles to each other. From this it is evident that the component *a* is acting against *b* to some extent, for the reason that *b*, *c*, and *d* together are equivalent to *b* and *a*. Subtracting *c* from *b*, we have left a horizontal velocity represented by the difference between *c* and *b*, which, combined with *d* (Fig. 20-b), will give a resultant identical with the resultant of *b* and *a*.

Algebraically, this is shown from the formula,

$$r^2 = a^2 + b^2 + 2ab \cos \theta.$$

If θ is greater than 90°, $\cos \theta$ is negative. Hence, when $2ab \cos \theta$ becomes greater than a^2, then the resultant r^2 becomes less than b^2; and when $2ab \cos \theta$ becomes greater than b^2, r^2 becomes less than a^2.

CHAPTER III.

CIRCULAR MOTION.

Motion of a body in a circle. It has been stated (p. 3) that a full description of the motion of a body involves:

1. Its direction of motion;
2. Its rate of motion.

A body moving in a circle may have uniform speed, but its motion differs from the rectilinear motions heretofore considered by constantly changing in direction. It is only necessary to recall the familiar geometrical distinctions between straight lines, broken lines and curved lines, in order to comprehend the difference between rectilinear and curvilinear motion,—the former constant in direction, and the latter constantly changing direction. Motion in a circle *at a uniform speed* is the simplest of curvilinear motions, because the *change of direction* takes place *at a uniform rate*.

The speed is found in the usual way, $v = d/t$. For instance, if a body is moving in the circumference of a circle with radius equal to r feet, and requires six seconds for completing one revolution, its velocity is $2\pi r/6$ feet per second. The distance traveled per second is $\frac{1}{6}$ of the circumference, or in angular measure 60°, as shown by dotted radius r (Fig. 21). If the radius had been twice as large, the body, moving at the same speed, would traverse an arc of only 30° in one second; and if the radius were half as large, the angular change would be 120°. The rate of angular change, therefore, depends upon the radius of the circle as well as upon the speed of the body; it is

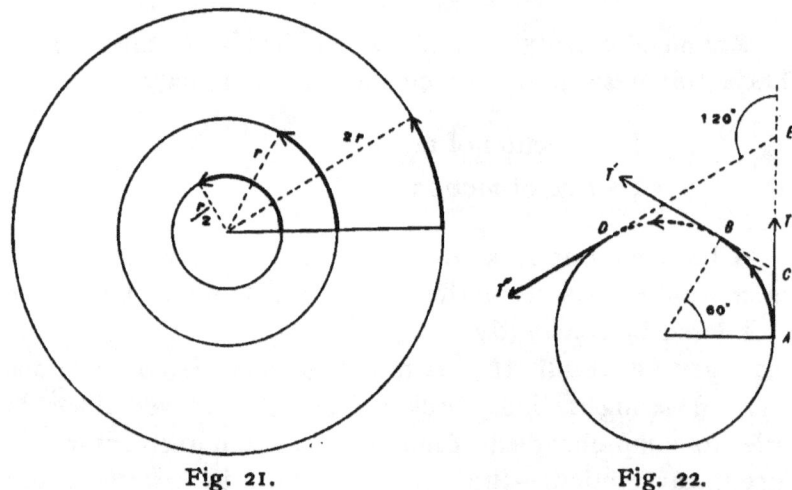

Fig. 21.　　　　　　Fig. 22.

directly proportional to one and inversely proportional to the other.

Properly, this rate of angular change should be determined from tangents rather than from the radii, as illustrated in Fig. 22. When the body moving in the circle is at point A, its motion for the instant is in the direction of the tangent AT; at point B its motion is in the direction of the tangent BT'. If the arc AB is 60°, the angle TCT' is also 60°. If the body moves in the circumference from A to B in one second, it not only has a velocity of $2\pi r/6$ feet per second, but the direction of its motion is also changing at the rate of 60° per second. At the end of 2 seconds the body will have covered a distance AD equal to $2AB$, and the total change of direction will be 120°.

The rate of angular change,

$$\frac{total\ change\ of\ direction}{time},$$

is frequently called the **Angular Velocity**. In this sense the word velocity does not have its usual meaning of distance traveled per unit of time, but is given a much broader significance, equivalent to the general expression "rate of change." This is a figurative use of the word that may be somewhat confusing to beginners. For example, "temperature velocity" would mean the rate at which the temperature is changing. In the same way, "angular velocity" means the rate at which the direction, (difference of direction, or angle), changes.

Examples:

1. *If the earth's radius is 4,000 miles, what is the velocity in miles per hour of a point on the equator, due to the earth's axial rotation? What is its angular velocity?*

2. *What lineal velocity and rate of angular change of the Lick Telescope (situated in latitude 37° 20' 25" North), is caused by the earth's axial rotation?*

3. *The moon is about 250,000 miles from the earth, and completes its revolution in about 28 days. What is its orbital velocity in miles per hour? What is its angular velocity?*

4. *The distance from the earth to the sun is 92,000,000 miles. If there were exactly 365 days in a year, what would be the orbital velocity of the earth's center and its angular velocity?*

5. *Two gear wheels having diameters of 4 inches and 12 inches, respectively, have fixed axes O and O'. If the larger wheel has 200 revolutions per minute, how many revolutions does it impart to the smaller wheel? What is the lineal velocity of each point on the circumference of the small wheel, and what is its angular velocity about O? What is the lineal velocity of each point on the circumference of the larger wheel, and what is its angular velocity about O'? Would the results have been the same if the motion had been transmitted from O' to O by belt instead of by gear?*

6. *A lathe is connected with a system of shafting, as illustrated in Fig. 23, the numbers indicating the diameters of the pulleys in*

28 THEORETICAL MECHANICS.

inches. *The speed of the main shaft is 200 revolutions per minute.*

(*a*) *Find the revolutions per minute, and the lineal velocity of the perimeter of each of the pulleys mentioned in the following table:*

PULLEY	ANGULAR VELOCITY REVOLUTIONS PER MINUTE	LINEAL VELOCITY FEET PER MINUTE
18 inch		
24 "		
9 "		
10 "		
5¼ "		
6½ "		
8 "		
9½ "		

(*b*) *Find the angular velocity of chuck, with and without back-gear. Tabulate results as follows:*

| POSITION OF BELT | Without Back Gear | With Back Gear | |
	Angular Velocity Cone and Chuck	Angular Velocity of Back Gear	Angular Velocity of Chuck
On 7¼ inch Pulley			
" 6 " "			
" 4½ " "			
" 3 " "			

CIRCULAR MOTION.

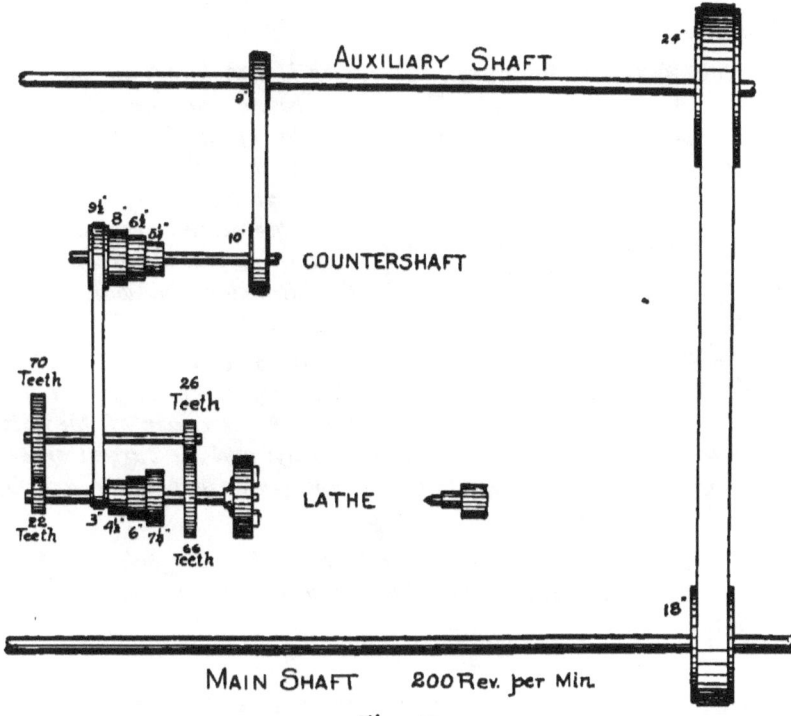

Fig. 23.

Composition of Circular and Rectilinear Motions. When a ball is thrown in any direction it usually possesses a rotary or whirling motion, in addition to and independently of its progressive motion or motion of translation. It is by his ability to produce and control these rotations that a skillful baseball pitcher causes a baseball to move in an erratic path,—"pitching curves," as it is called. If the ball is moving in a straight line and rotating around an imaginary axis passing through its center, the actual motion, relatively to the ground, of any point on its surface at any instant is the resultant of a circular or rotational motion and a rectilinear motion or translation. A simple case is assumed in the example on the following page.

30 THEORETICAL MECHANICS.

Example:

The Baseball Problem. *A baseball, 2.75 inches in diameter, is thrown horizontally with a velocity of 80 feet per second, and at the same time is made to rotate at the rate of 30 revolutions per second around a horizontal axis at right angles to the direction in which it is thrown.*

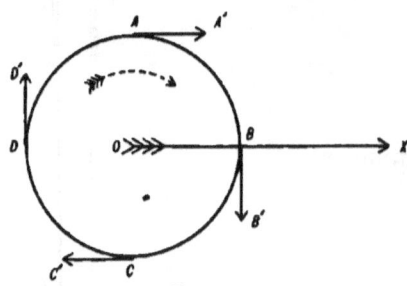

Fig. 24.

In Fig. 23, let the circle ABCD represent a vertical section through the center of the ball. The line OX represents the horizontal velocity, and the curved arrow indicates a rotation around an axis imagined to pass through O, perpendicular to the plane of the paper.

(a) *Find the resulant velocity of each of the points* A, B, C *and* D, *at the instant when they are in the position shown in the figure.*

DISCUSSION:—The entire mass of the ball, including the points A, B, C and D, moves horizontally with the same velocity, OX, as the center, and hence OX is one of the components for each of these points.

The other component, due to the rotation of the ball, is different for each point of the ball; for all points on the circumference of the circle $ABCD$, the rate or magnitude of this rotational component is the same, but no two points are moving in the same direction at the given instant. When C is at the lowest position on the ball, the rotation gives it a motion for the instant in the direction of the tangential arrow CC'; the rotational component of point D at that instant would be vertically upward; of A horizontally to the right; and of B vertically downward.

It is important that the student should get a clear conception of what is meant by these tangential components. True, the points move in the circle and not along the tangential lines. They follow the circular path, however, merely because they are constrained to do so. It is not difficult to see that, as the ball revolves and C moves from its present position to that now occupied by D, if at the instant when it reaches the latter point it is freed from this constraint (which would pull it around in the circle towards A), it will move instead along the tangent DD',—like a drop of water similarly freed and flying from a grindstone.

Now, while the points A, B, C and D are not thus freed from the ball, still their respective motions are for the instant in the directions of the tangents. Imagine the circle made up of an infinite number of very small straight lines—a regular polygon of an infinite number of sides—and the same conclusion is reached. The infinitesimal line constituting any part of the circumference will have the direction of the tangent at that point. The rotational component of D is vertically upward for only an infinitesimal fraction of a second, and then it assumes a different direction. But, if we wish to combine this rotational component with the horizontal component OX, we must represent it graphically on the same scale. Hence, if OX represents one of the velocities of D in feet per second, then the other, in direction DD', must also be represented in feet per second. The point D does not continue to move in direction DD' for a full second, but if it did it would cover the distance DD', on the same scale as OX. When we say that a train has a velocity of 30 miles per hour, we do not insist that it shall move the full hour, but we can represent its motion graphically as if it did; so we say of the ball, if point D moves a very small distance in direction DD' in a very small interval of time, at the same rate it would in a full second move the distance DD'.

Hence, the motion of D, relatively to the ground at the given instant, is the resultant of the two components OX and DD' at right angles to each other. And this will determine the direction of the resultant as well as its magnitude. The two components for B are also at right angles to each other; those for A and C are parallel.

Notice that the ball is represented in Fig. 24 on a scale much larger than the scale of velocities. If it makes 30 rotations per second it thereby causes the points A, B, C, and D to move in a second through a distance equal to 30 times the circumference shown in the figure,—which, as it is represented in the diagram, is many times greater than the line DD'. It is obvious that the velocities had to be represented on a greatly reduced scale.

(*b*) *Find the magnitude and direction of the resultant velocity of a point on the ball between* A *and* B, *22° from* B.

The motion of a carriage wheel is not unlike that of the ball in the preceding example, except that the rate of rotation of the wheel is not independent of the velocity of the vehicle. A baseball may revolve fast or slowly, without regard to the velocity with which it is thrown, but the rate at which a carriage wheel revolves necessarily bears a fixed relation to the velocity of the vehicle. In one revolution the wheel measures the length of its circumference on the ground. Hence, the rotational component of any point on the perimeter of the wheel is exactly equal to the rectilinear component parallel to the ground. Otherwise, the general conditions are the same as those assumed for the baseball in the preceding example.

Examples:

 1. *We know that that point on the tire of a wheel momentarily touching the ground is at rest relatively to the ground*; otherwise there would be slipping. Prove this by the composition of the rectilinear and rotational components.*

 2. *If v is the velocity of the vehicle, prove that the highest point on the wheel has a resultant velocity of 2v in the same direction.*

 3. *A carriage is traveling at the rate of 9 miles per hour.*

 (*a*) *Find the resultant velocity relatively to the ground of a point on the front side of one of the wheels 40° from the ground.*

 (*b*) *If the diameter of the front wheels is 3½ feet and of the hind wheels 4 feet, find the angular velocity of each.*

Composition of two Circular Motions. Every point on the earth's surface is subject to at least two motions, one due to the rotation of the earth on its axis, and the other due to its orbital motion or revolution around the sun.

---

* If there is any doubt of this in the mind of the student, try the experiment of tying a piece of white cloth, or marking with a piece of chalk, around a bicycle tire. It will be seen that when the bicycle is ridden, the cloth or chalk mark comes to rest as it reaches the ground.

CIRCULAR MOTION. 33

Hence, the resultant motion of a point A (Fig. 25) is different from that of a point B.

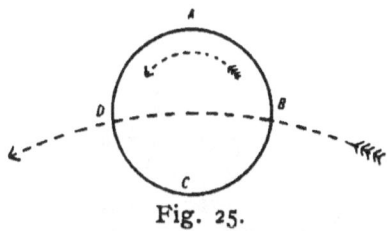

Fig. 25.

As the moon revolves around the earth, however, it does not possess an additional *independent* motion around its own axis. Its motion in this respect is analogous to the motion of a circular disc fastened to a stick, as in Fig. 26. If this whole device is made to revolve around point O of the stick in the direction of the arrow, a person standing beyond A would see successively points A, B, C, D. To him, therefore, there would be an apparent rotation of the disc around its center, but this apparent rotation is really controlled by the revolution about O.

This does not include the consideration that the moon is carried with the earth in the motion of the latter around the sun, and hence has a double motion after all. In fact, if the moon had an independent axial rotation each point in it would then be subject to three simultaneous, circular motions.

Since the earth's axis is inclined to its orbit, and since the moon's orbit around the earth is inclined to the earth's orbit, the determination of the resultants of these motions is not

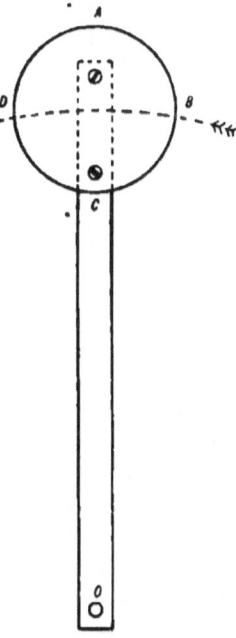

Fig. 26.

simple enough for our purposes of explanation. The same ideas, however, are illustrated in modified form in the following

Examples:

1. *A circular disc 6 inches in diameter is fastened to a stick. This device is made to revolve 25 times per minute around an axis which passes through the stick at a point 18 inches from the center of the disc. Find the tangential velocity of the center of the disc and of each of four points situated as A, B, C and D, Fig. 26, (line BD being perpendicular to AC.)*

2. *Now imagine the screws removed and the disc no longer fixed rigidly to the stick, but free to rotate around a pin P at its center, as in Fig. 27. If the disc rotates 70 times per minute around P, in addition to its revolution around O, what will be the resultant velocity of each of the points P, A, B, and C?*

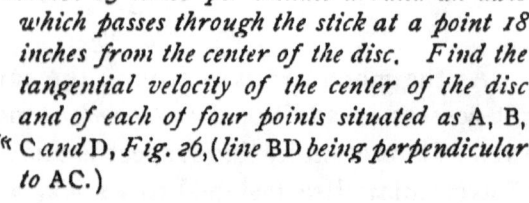

Fig. 27.

This is an arrangement that would be quite analogous to the axial and orbital motions of the earth, if the earth's axis were perpendicular to the plane of its orbit.

Comparing this example with the baseball problem will show how the composition of two circular motions differs from the composition of a circular with a rectilinear motion. The revolution of the disc (Fig. 27) in a circular path around *O* gives point *A* a greater component than point *C*, but if the disc were moving in a straight line, like the baseball in Fig. 27, this rectilinear component would be the same for all points.

In the baseball problem every point on the perimeter of the section shown in the figure had the same pair of components, the only difference for different points being in the angle between the two components; in Fig. 27, not only is the angle different for each

CIRCULAR MOTION.   35

pair of components, but only one of the components is the same for all points.

3. *If the earth's axis were perpendicular to the plane of its orbit, what would be the resultant velocity of a point on the equator at midnight, at noonday, and at 7 o'clock p. m.?*

Assume: One year equal to 365 days; diameter of the earth at the equator equal to 8,000 miles; distance of the center of the earth from the sun equal to 92,000,000 miles; that the earth revolves around the sun from west to east and has its axial rotation in the same direction.

4. *If the moon's orbit around the earth were in the same plane as the earth's orbit, what would be the resultant velocity of the center of the moon at "new moon"?*

Assume that the moon completes its circuit around the earth in 28 days, and that the motion is from west to east, like the orbital motion of the earth.

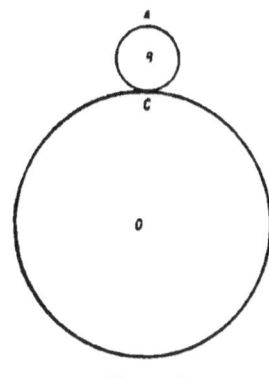

Fig. 28.

5. *A circle 6 inches in diameter rolls around a larger circle, diameter 24 inches, at the rate of 50 revolutions per minute. B is the center of the small circle, and A and C points on its circumference. What are the lineal and angular velocities of B? What is the resultant lineal velocity of A and of C?*

This problem bears the same relation to the three examples preceding that the carriage wheel motion bears to the baseball problem; the axial rotation of the earth is not in a simple ratio to its orbital motion, and is independent of it, but in this problem the two motions of the small circle are inter-dependent, if there is no slipping.

Cycloids. If a piece of chalk or other marker be fastened at the perimeter of a disc, and the latter be rolled along the floor with the disc against the wall, the marker will trace a line like the curve shown in Fig. 29. This curve is called a **Cycloid.**

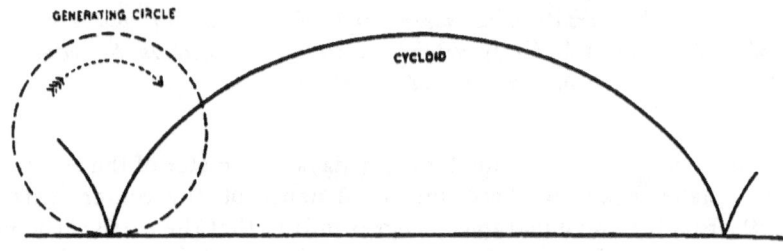

Fig. 29.

Every point on a carriage wheel describes such a path. An inspection of this curve discloses the fact, already referred to, that every point on the perimeter of the disc or wheel comes to rest as it reaches the ground. Approaching the ground, it is moving almost vertically downward, and as soon as it begins to leave the ground its motion is upward. Having turned back abruptly, as the diagram shows, it must have come to rest at the turning point.

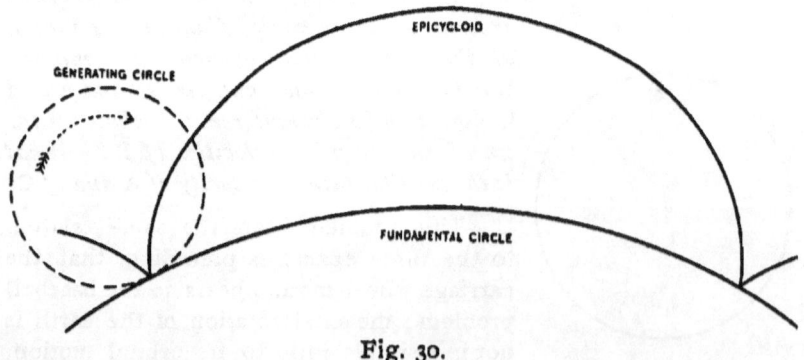

Fig. 30.

The cycloid generated by a point on the circumference of a circle as it rolls along a straight line is only one of a class of similar curves that are used in mechanics, especially in designing gear teeth. If a circle, instead of rolling along a straight line, be rolled on the circumference of another circle, as in Fig. 30, it will generate a curve called an **Epicycloid**. If it be rolled on the inside of a ring in the same direction as before, its rotations around its own axis will be in the opposite direction, and the curve generated in this case is called a **Hypocycloid**.

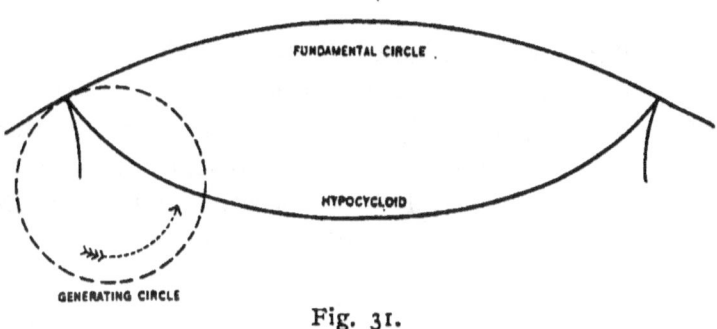

Fig. 31.

For methods of constructing these curves geometrically see Kent's Mechanical Engineers' Pocket Book, pp. 49 and 50, or Appleton's Cyclopædia of Drawing and Design, pp. 303 and 304.

CHAPTER IV.

ACCELERATION.

A perfectly uniform rate of motion is seldom realized. The movements of a clock or a line of shafting are generally assumed to be uniform, but are really subject to many fluctuations. An electric car moves with increasing velocity up to the desired speed, maintains an apparently uniform velocity for a time, and again assumes a distinctly variable velocity before stopping. A falling body gains velocity continuously; and if hurled upward, loses velocity steadily until it again starts downward.

The subject of **acceleration** deals with all cases of variable velocity. When the rate of motion changes irregularly or spasmodically it generally involves too many complexities for treatment by arithmetical or algebraic methods. Hence, our study of acceleration will be limited to motions that change uniformly,—called uniformly accelerated motions.

It should be remembered that changes of direction, as exemplified by motion in a circle, need not affect the rate of a motion. Unless otherwise specified, a motion is always assumed to be in a straight line.

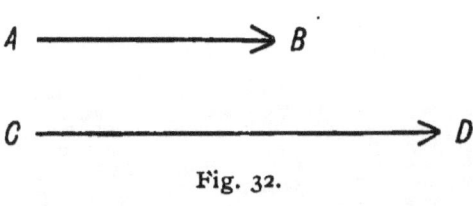

Fig. 32.

Suppose that a body is observed at a given instant to have a velocity of 10 miles per hour, represented by the line AB, Fig. 32.

ACCELERATION.   39

If, after a few minutes—say three minutes, it is found to have a velocity of 16 miles per hour, represented by the line *CD*, the question arises: Was this additional velocity caused by some sudden impulse, or was it a steady increase distributed equally over the entire interval of three minutes? If we take the latter supposition, and divide the total increase of velocity (6 miles per hour) by the time during which the increase took place (3 minutes), the quotient will represent the amount of velocity (in miles per hour) by which the motion of the body was increased during each of the three minutes, or its acceleration in miles per hour, per minute. At the end of the first minute the velocity must have been 12 miles per hour; at the end of the second minute, 14 miles per hour; etc. At the end of 1½ minutes it was 13 miles per hour; at the end of 2¼ minutes it was 14.5 miles per hour; etc. The rate of motion is not the same at any two successive instants, and the body does not travel through any appreciable distance at a fixed rate.

Some of these things can be shown graphically. Draw a horizontal base-line *OX* and a vertical base-line *OY*. Let intervals of time be represented on a convenient horizontal

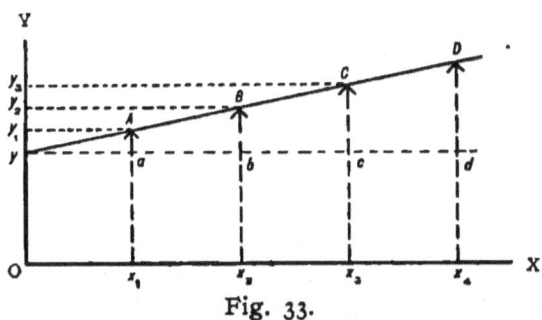
Fig. 33.

scale, say 1 inch = 1 minute, and use a vertical arrow to indicate the velocity at any instant, on a scale of 1 inch = velocity of 10 miles per hour. (The dimensions of Fig. 33

are reduced one-half.) Oy will represent the velocity as first observed, and Oy_3 will represent the velocity as observed after the lapse of 3 minutes. Ox_3 will represent the lapse of 3 minutes. The line x_3C, parallel to OY, contains all points 3 inches from the Y-axis; the line y_3C, parallel to OX, contains all points 1.6 inches from the X-axis. The intersection of x_3C and y_3C locates a point which pictures the two things, viz.,—that at the end of 3 minutes, Ox_3, the velocity of the body is 16 miles per hour, x_3C. To show that the rate of motion has increased uniformly we have only to connect y and C by a straight line. The upward inclination of yC shows that the rate of motion is increasing, and the steepness of the gradient pictures the rapidity of the change. The total increase of velocity for the 3 minutes is cC.

Supposing that the rate of motion of this body continues to increase at the same rate as observed for the interval of 3 minutes, can we calculate the velocity it would have at any subsequent time? If the rate of increase—2 miles per hour, per minute—continues, then the velocity at the end of the fourth minute from the beginning will be $10 + 4 \times 2$, or 18 miles per hour, which would be represented by the line x_4D in the diagram.

Examples:

    1. *Taking a point* x_1 (*Fig. 33*) *so that* Ox_1 *will represent one minute, what will be represented by the perpendicular* x_1A*? What will* aA *represent? How could we find from the diagram the velocity of the body at the end of 1.7 minutes? At what time after starting would the velocity be 15.3 miles per hour? What does* dD *represent?*

    2. *A body starting from rest has its velocity increased gradually and uniformly until at the end of one second its velocity is found to be 10 feet per second.*

ACCELERATION.

(a) What has been its average velocity meanwhile?
(b) Suppose its velocity to increase always at the same rate; what would be its velocity at the end of the second second?
(c) What will have been its average velocity during the first two seconds?
(d) What average velocity during second second?
(e) What average velocity during first three seconds?
(f) What average velocity during third second?

3. (a) What distance will be passed over by this body during first second?
(b) During the first two seconds?
(c) During the second second?
(d) During the first three seconds?
(e) During the third second?

4. A train of cars leaving a station acquires velocity gradually and uniformly until at the end of three minutes it is moving at the rate of 30 miles per hour.

(a) What velocity in miles per hour has been added each minute?
(b) What velocity in feet per minute has been added each minute?
(c) What velocity in feet per minute has been added each second?
(d) What velocity in feet per second has been added each second?

5. A body starting from rest is accelerated uniformly. At the end of 5 seconds it has a velocity of 100 feet per second.

(a) What has been its average velocity during the five seconds?
(b) How far has it traveled?
(c) What was its velocity at the end of the first second?
(d) What is its acceleration?
(e) How far did it travel during the first second?
(f) What was its velocity at the end of the second second?
(g) How far did it travel during first two seconds?
(h) How far did it travel during second second?
(i) How far did it travel during the first ten seconds?
(j) How far did it travel during the fourteenth second?

Definition of Acceleration. By the acceleration of a body we mean the rate at which its rate of motion, or velocity, changes.

In the first illustration of this idea at the beginning of this chapter, the velocity changed from 10 miles per hour to 16 miles per hour in 3 minutes,—the change taking place at the rate of 2 miles per hour for each minute; and hence, according to the definition, the acceleration was 2 miles per hour, per minute.

In example 2, p. 40, the acceleration was 10 feet per second, per second; in exercise 4 it was 10 miles per hour, per minute, or $\frac{1}{6}$ mile per hour, per second, (and might have been expressed in any other units of time and distance, as feet per minute, per minute; feet per minute, per second; feet per second, per second; inches per second, per second, etc.)

Formulæ for Uniformly Accelerated Motion. From the preceding examples and statements we can generalize to the following relations:

(i) If a body starting from rest is subject to a uniform acceleration, a, its velocity, v, at the end of any time, t, will be
$$v = at. \qquad (4)$$

(ii) If the body does not start from a condition of rest its motion may be either accelerated or retarded, the latter being regarded as a negative acceleration. If its initial velocity is v_1, the velocity at the end of any given interval, t, will be
$$v = v_1 + at. \qquad (5)$$

(iii) To find the distance traveled in any time by any body, no matter how it is moving, we multiply the average

ACCELERATION.

velocity by the time. If the body starts from rest and is subject to a uniform acceleration a, the velocity at the *end* of time t, is at; the average velocity *during* t is $at/2$. The distance traveled is

$$d = \frac{at}{2} \times t = \frac{at^2}{2}. \tag{6}$$

If, instead of starting from rest, the motion changes uniformly from an initial velocity v_1, to a final velocity v_2, then the distance traveled meanwhile will be

$$d = \frac{v_1 + v_2}{2} \times t.$$

But v_2 can be expressed in terms of v_1 and the acceleration, thus:

$$v_2 = v_1 + at.$$

Whence

$$d = \frac{v_1 + v_1 + at}{2} \times t = \left(\frac{2v_1 + at}{2}\right) t = v_1 t + \frac{at^2}{2}. \tag{7}$$

Examples:

 1. *A body starting from rest moves with a uniformly accelerated motion. When it has passed over a distance of 100 feet its velocity is found to be 50 feet per second.*
   (a) *What has been its average velocity meanwhile?*
   (b) *How long has it taken the body to travel the 100 feet?*
   (c) *What is its acceleration?*
   (d) *What was its velocity at the end of 2.6 seconds?*
   (e) *What will be its velocity at the end of 1 minute, and how far will it have traveled meanwhile?*

 2. *A body starting with a velocity of 100 feet per second loses its velocity gradually and uniformly and comes to rest in 6 seconds.*
   (a) *What is its acceleration?*
   (b) *How far will it have traveled before coming to rest?*
   (c) *How far will it have traveled at the end of 3 seconds?*

3. *A body starting with a velocity of 30 feet per second is accelerated uniformly at the rate of 8 feet per second, per second.*
    (a) *What will be its velocity at the end of 5 seconds?*
    (b) *How far will it have traveled meanwhile?*
    (c) *How far will it travel during the thirteenth second?*

4. *A body starting with a velocity of 50 feet per second is accelerated uniformly. At the end of 5 seconds it has a velocity of 200 feet per second.*
    (a) *What is its acceleration?*
    (b) *How far does it travel during the second, third and fourth seconds inclusive?*

5. *A body starting with a velocity of 50 feet per second is accelerated uniformly. At the end of 10 seconds it is found to have traveled 2,000 feet.*
    (a) *What has been its average velocity?*
    (b) *What was its final velocity?*
    (c) *What was its acceleration?*

6. *A body starting with a velocity of 50 feet per second loses velocity at a uniform rate. At the end of 3 seconds, it has a velocity of 20 feet per second.*
    (a) *What was its acceleration?*
    (b) *How long before it will come to rest?*
    (c) *How far will it move before coming to rest?*

7. *If a body starts from rest and moves through the distance* d *while it is acquiring a velocity* v, *prove that*

$$d = \frac{v^2}{2a}, \qquad (8)$$

where a *is the acceleration.*

Acceleration of Gravity. The acceleration of gravity is uniform and equal to about 32.2 feet per second, per second. It is usually represented by the letter *g*. Under the influence of the attraction that exists between the general mass of the earth and objects on its surface, a falling body

ACCELERATION.

gains velocity at the rate specified; a body hurled upward loses velocity at the same rate.

The value of g is different for different parts of the earth's surface. It is greater at the poles than at the equator and is subject to other irregularities.

Examples:

   1. A body falling from a certain height requires 5 seconds to reach the ground. With what velocity does it strike? What was its velocity at the end of 1.5 seconds? At the end of 3.75 seconds? From what height was it dropped? What part of the distance was covered in each of the five seconds? What part was covered in the first 1.7 seconds? In the first 4.4 seconds? During the interval between the end of the second second and the end of 3.8 seconds?

   2. A falling body strikes the ground with a velocity of 193.2 feet per second. From what height was it dropped?

   3. A body is dropped from a height of 788.9 feet. How long before it will strike the ground? With what velocity will it strike?

   4. A body is hurled downward with a velocity of 100 feet per second. What will be its velocity at the end of 3 seconds? How far will it have traveled? How far will it have traveled at the end of 5 seconds? What part of this distance will have been covered during each of the 5 seconds? During second and third seconds, inclusive?

   5. A body being hurled vertically downward strikes the ground at the end of 7 seconds with a velocity of 525.4 feet per second. With what velocity was it hurled? From what height was it hurled?

   6. A body is hurled downward with a velocity of 20 feet per second from a height of 1810 feet. How long before it will strike the ground?

   7. A body is projected vertically upward with a velocity of 1000 feet per second. How long will it continue to rise? How far will it rise? With what velocity will it strike the ground on returning?

8. A man reaching from the top of a tower 644 feet high, throws a ball vertically upward with a velocity of 96.6 feet per second. How long before it will reach the ground?

9. If the acceleration of gravity is g feet per second, per second, express the following relations:

(*i*)   v *in terms of* g *and* t ;
(*ii*) h *in terms of* g *and* t ;
(*iii*) h *in terms of* g *and* v.

The body is assumed to start from rest ; v is its velocity at the end of t seconds, and h is the distance it falls meanwhile.

10. What is the value of g expressed in cm. per second, per second?

Use of Co-ordinate Axes. The method of locating points with reference to two lines or axes, employed in Fig. 33, p. 39, and elsewhere in this book, is used very widely in various branches of mathematics and science. The two reference lines may be drawn at any angle with each other, although they are usually perpendicular, one being drawn horizontally and the other vertically. Their intersection is called the **origin**. The horizontal line is called the **X-axis**, or axis of abscissas, and the vertical line is the **Y-axis**, or axis of ordinates.

The familiar geographical method of locating places by means of latitude and longitude assumes two such reference lines—the terrestrial equator and the arbitrary line through Greenwich called the zero meridian. What do we mean, for example, when we say that a place is situated in latitude 30° N., longitude 80° W.? The thirtieth parallel of north latitude is the location or *locus* of all points 30° north of the equator, and the eightieth meridian west is the *locus* of all points 80° west of the zero or Greenwich meridian. The intersection of these two lines or *loci* is the point described.

Tracts of public land are located in the same manner. In the central portions of California, for instance, arbitrary east-and-west and north-and-south lines, intersecting at the summit of Mount Diablo and called the Mount Diablo Base and Meridian, have been selected, and a piece of land is described as being located so many townships north and east of the Mount Diablo Base and Meridian,—or N. and W., or S. and E., or S. and W., as the case might be.

In Fig. 33, p. 39, by locating points A, B, C, etc., with reference to two axes at right angles to each other, we represented the gradual changing of a velocity during several seconds, in accordance with the law expressed in formula 5, p. 42, $v = v_1 + a t$. Now it is true generally that any relation between two quantities can be thus represented as a line, either straight or curved, and the character of this line will picture the nature of the relation between the two quantities.

With reference to two such axes every simple proportion and every simple equation between two quantities can be represented by a straight line. The weight of a given volume of anything depends upon its density, or $W = V \times d$; the electrical resistance of a wire is proportional to its length; the bending of a beam is directly proportional to the load upon it. Each of these relations could be represented by a straight line.

More complex relations give rise to curves of various sorts or even to broken lines. Some quadratic equations produce ellipses, others circles, parabolas, etc. Temperature curves and barometric curves, showing the gradual rise and fall of the thermometer or barometer throughout a day or other interval of time, are usually very irregular. A more interesting case is the diagram on the indicator card from which the horse-power of an engine is determined, the relation in this case being between the pressure of steam in the cylinder and the distance of the piston from either end of the stroke. Fig. 34 shows

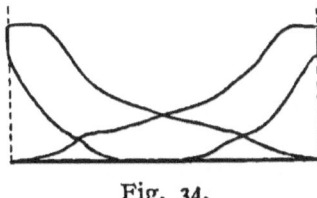

Fig. 34.

such a diagram* taken from an engine of 12-inch stroke under a boiler pressure of 75 pounds per square inch, so that the total horizontal width of the diagram corresponds to 12 inches of stroke and the highest point on the curves corresponds to about 75 pounds pressure per square inch.

In every case, whether straight or curved, the line pictures a relation between two quantities—velocity and time; temperature and

---

*Taken from a 9.5-inch x 12-inch Armington & Sims Standard Automatic Cut-off Engine running 200 revolutions per minute, under boiler pressure of 75 pounds per square inch, indicating 24.77 H. P.

time; weight and density; etc.,—one of the quantities being a function of the other, or the changes in one accompanying changes in the other.

Graphical Representation of Distance Traveled by an Accelerated Body. To the lines and points in a diagram constructed upon co-ordinate axes various meanings may be attached, depending upon the idea to be illustrated. Sometimes, furthermore, the usefulness of the figure depends upon the interpretation of the area included within the various lines and axes as much as upon the lines themselves. For example, in the indicator diagram from which the power and workings of a steam engine are investigated, the horse-power is determined from the area within the curves.

Reverting to Fig. 33, p. 39, which is here reprinted as Fig. 35, it can be shown that the area Ox_3Cy represents the distance traveled by the body referred to. Oy represents the initial velocity, and Ox_3 represents an interval of time—3 minutes. If the velocity had continued equal to Oy (10 miles per hour) throughout the 3 minutes,—that is, if there had been no acceleration,—the distance traveled during that time would have been $v \times t = \frac{10 \times 3}{60} = \frac{1}{2}$ mile. In the diagram v is represented by Oy and t is represented by Ox_3. The product of Oy and Ox_3 is the area of the parallelogram Ox_3cy, whence it appears that the distance traveled by any body moving with a uniform velocity can be represented by the area of a rectangle of which two adjacent sides represent the velocity and time, respectively.

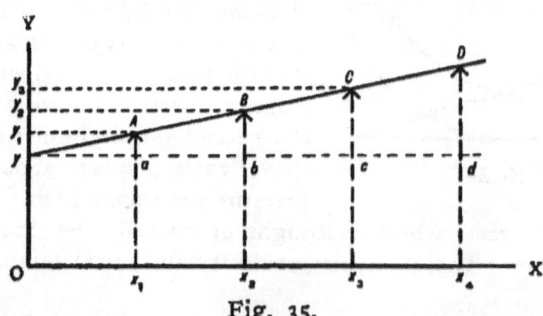

Fig. 35.

What in the case of accelerated motion? If the body under consideration, instead of moving uniformly at the rate of 10 miles

per hour, has an acceleration of 2 miles per hour, per minute, increasing from Oy in the diagram to Oy_3 or x_3C in the course of 3 minutes it is apparent that the area Ox_3cy will represent only a part of the total distance. Study the diagram by comparison with the formula $d = v_1 t + \frac{a t^2}{2}$. It has just been shown the term $v_1 t$ is represented by the area Ox_3cy, which indicates what the distance would have been if the velocity had remained constant. The term $\frac{a t^2}{2}$ is represented by the area of the triangle ycC,* and indicates the additional distance covered on account of the increased motion. The total area Ox_3Cy, therefore, represents the sum of the two, or the distance traveled in 3 minutes by a body starting with a velocity of 10 miles an hour and subject to an acceleration of 2 miles per hour, per minute.

This method could be used even for variable accelerations, thereby avoiding many complex computations. No matter how erratic the changes from velocity Oy to velocity x_3C, if they can be put into a diagram as in Fig. 36, we can easily determine the area Oxy_1y,

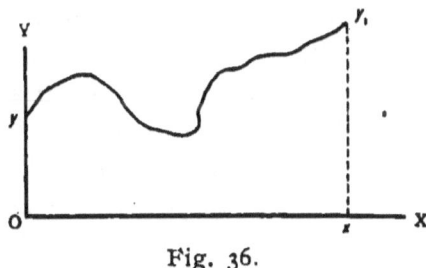

Fig. 36.

which represents the distance traveled. This area divided by the time Ox, will give the average velocity or the average of all the possible vertical co-ordinates.

[A simple plan for determining the irregular area Oxy_1y would be to cut out the figure from a piece of cardboard and compare the weight of this irregular piece of cardboard with the weight of a regular piece, or known area, of the same material.]

---

*$cC = at$ and $yc = t$; whence the area of the triangle is equal to $\frac{yc \times cC}{2}$, or $\frac{at^2}{2}$, as stated.

PROJECTILES.

The path of a free projectile, undisturbed except by the action of gravity, is always a curve. Whatever the velocity or angle of projection, this curve always possesses certain characteristic features or mathematical relations peculiar to a class of curves called **parabolas**. A projectile is always subject to two component motions, one of which (the velocity of projection) is constant in both magnitude and direction, while the other (the gravitational component) is constant in direction but of variable magnitude. If both components were entirely uniform velocities the resultant path would be a straight line instead of a curve, as in Figs. 3 and 4, pp. 9 and 10. If it were not for the action of gravity, a free and unobstructed body hurled in any direction—horizontally, obliquely, or vertically,—would move on forever in a straight line with a uniform velocity. The effect of gravitation, however, adds this variable downward component, the value of which at any instant can be determined by the law of falling bodies. The motion of a projectile at any instant is, therefore, the resultant of two components,—one a uniform velocity in the direction of projection, and the other a uniformly accelerated velocity vertically downward.

For instance, suppose a ball is rolled along a platform OA, Fig. 37, with a uniform velocity, say 40 feet per second. If the platform remains at rest, the position of the ball at the end of successive seconds will be at points 1, 2, 3, etc. But if the platform is allowed to fall freely, at the end of 1 second it will be in position O_1A_1, 16.1 feet below OA, and the ball will be at 1'. At the end of 2 seconds the ball will be to 2', 64.4 feet below 2, etc.

ACCELERATION.   51

It should be noted that this diagram (Fig. 37) is very different from some of the graphical representations previously used; in this case the horizontal and vertical magnitudes represent distances and not velocities, and the curved line is not the resultant velocity, but the actual path traversed. The resultant velocity at any instant is a tangent to the curve at the point where the body happens to be at that instant. To compute this resultant we must combine

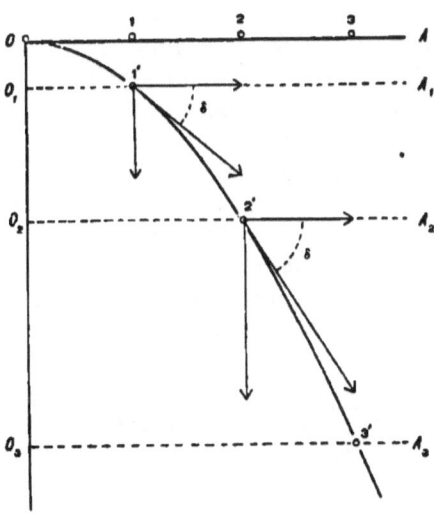

Fig. 37.

the fixed horizontal component and the computed vertical component for the given instant. These velocities may be represented graphically on any scale desired, independently of the scale of distances used in constructing the path. At point O, for example, the horizontal component is 40 feet per second and the vertical component is zero; the resultant is 40 feet per second horizontally. At the very next instant—the merest fraction of a second—it is something different. At point 1' the horizontal component is 40 feet per second and the vertical component is 32.2 feet per second; the resultant is $\sqrt{40^2 + 32.2^2}$, at an angle $\delta = \tan^{-1} 32.2 \div 40$.* At point 3' the horizontal component is still 40 feet per second and the vertical component is 96.6 feet per second; the resultant is $\sqrt{40^2 + 96.6^2}$, at an angle $\delta = \tan^{-1} 96.6 \div 40$.

The resultant of two uniform components is constant in both magnitude and direction; uniform motion in a circle is constant in rate but continually changes direction; the motion of a projectile changes in both magnitude and direction.

Another interesting contrast is had by comparing the motion of a projectile with the resultant motion of a point on the perimeter of a carriage wheel describing a cycloidal path, as in Fig. 29. A point on the carriage wheel is subject to two component motions, both of which remain constant in magnitude, while in the matter of direction one component remains always parallel to the ground and the other (the tangential component) continually changes direction as the wheel revolves. The point on the wheel, therefore, has both components constant in magnitude and one changing in direction; a projectile has both components constant in direction, and one changing in magnitude.

In the baseball problem, p. 30, we assumed that the ball maintained a horizontal direction like the carriage wheel. If we allow for the action of gravity, then it will assume a parabolic path like any projectile, and the motion of any point in it will be the resultant of *three* components: (1) the velocity of projection, constant in all respects; (2) the tangential component due to the rotation of the ball, constant only in magnitude; (3) the gravitational component, constant only in direction.

Examples:

1. (*a*) *Construct a diagram of the path of a projectile hurled horizontally with a velocity of 100 feet per second from a height of 579.6 feet.*

INSTRUCTIONS: For graphical representation use a scale of 1 inch = 100 feet. First prepare a table of distances, such as the

---
*$\tan^{-1} 32.2 \div 40$ means "the angle whose tangent is $32.2 \div 40$."

following, in which the time refers to the number of seconds after starting, and the corresponding horizontal and vertical distances traveled are reduced to scale.

Time.	Horizontal Distance.		Vertical Distance.	
		[Reduced to Scale.]		[Reduced to Scale.]
$t = 0.5$ seconds	50 feet	0.5 inches	4.0 feet	0.040 inches
1.0 "	100 "	1.0 "	16.1 "	0.161 "
1.5 "	150 "	1.5 "	36.2 "	0.362 "
2.0 "	200 "	2.0 "	64.4 "	0.644 "
2.5 "	250 "	2.5 "	100.6 "	1.006 "
3.0 "	300 "	3.0 "	144.9 "	1.449 "
4.0 "	400 "	4.0 "	257.6 "	2.576 "
5.0 "	500 "	5.0 "	402.5 "	4.025 "
6.0 "	600 "	6.0 "	579.6 "	5.796 "

Prepare a cross-section sheet divided into one-inch squares with extra vertical lines at half-inch intervals where required by table. Beginning at upper left-hand corner as the origin, locate points determined by values reduced to scale in table, and connect successive points, starting from origin. In preparing diagram use a sharp, hard pencil.

(*b*) *Compute the resultant velocity of this body at the instant of striking the ground, giving the direction of the resultant as well as its magnitude.*

(*c*) *What was its resultant velocity at the end of the third second?*

2. *A body is projected with a velocity of 20 feet per second at an angle of 35° above the horizon. Construct a diagram of its path.*

INSTRUCTIONS: Use a scale of 1 inch = 5 feet. Tabulate values of distances traveled in direction of projection, and the falling distances, for successive periods of time changing by 1/10 second, after manner of last example. Trace the path up to the end of 1.5 seconds. In ruling the cross-section sheet draw the usual vertical base line, but construct the second reference line at the proper angle of 35° above the horizon. On the latter, measure off distances of 0.4 inch (representing distance of 2 feet traveled in that direction during each 0.1 second), and through each of these points draw a vertical line. In locating points determined by the values in table, count the uniform values along the inclined reference line, and from each of the successive points marked on that line, measure the falling distance for the time interval referred to.

The path should resemble the curve in Fig. 38.

3. Prove:

(*i*) *A body hurled horizontally from any altitude reaches the ground in the same time as if it had been merely dropped and allowed to fall vertically.*

(*ii*) *A body projected at any angle reaches the ground in the same time as if it had been projected vertically with a velocity equal to the vertical component of the actual velocity of projection.*

Elevation and Range of a Projectile. The angle of projection above the horizon is called the **Elevation**. The horizontal distance OA, Fig. 38, is called the **Range**. The

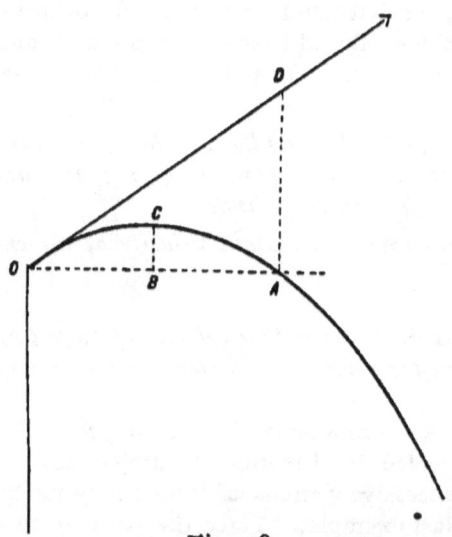

Fig. 38.

highest point of the path is called its **Vertex**, of which BC is the height.

The Range and Height of Vertex are easily determined by computation. For this purpose resolve the velocity of projection into its horizontal and vertical components. A

ACCELERATION.

body projected with a velocity v_p, at an angle β, will rise to

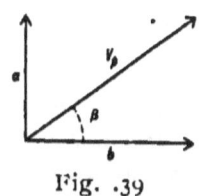

Fig. 39

the same vertical height and in the same time as if it were thrown vertically upward with a velocity equal to component a. Now, we have already learned how to find the height to which a body will rise if hurled upward with a given velocity, and the time that will elapse before it returns to the ground. This time multiplied by the horizontal component will give the range.

For example, the Range and Height of vertex in example 2, p. 53, would be determined as follows:

$$v_p = 20$$
$$a = v_p \sin 35° = 20 \times .5736 = 11.472$$
$$b = v_p \cos 35° = 20 \times .8192 = 16.384$$

An upward vertical velocity suffers an acceleration of —32.2 feet per second, per second on account of gravity, whence this body will reach its highest point in

$$t = \frac{11.472}{32.2} = 0.356 \text{ second,}$$

at an altitude of

$$h = \frac{11.472}{2} \times .356 = 2.042 \text{ feet.}$$

To find the range, R, multiply the horizontal component of v_p by $2t$, the time that elapses while the body moves to the highest point and back to the horizontal.

$$R = 2 \times .356 \times 16.384 = 11.665 \text{ feet.}$$

Examples:

   1. *From your diagram constructed for example 2, p. 53, determine the Range and Height of Vertex by scale, and compare with the above results.*

   2. *A body is projected with a velocity of 20 feet per second at an elevation of 45°. Construct a diagram of its path up to the end of*

1.5 seconds. Determine its height of vertex and range by scale and by computation.

3. Do likewise for a projectile hurled with a velocity of 21 miles per hour at an elevation of 55°.

4. A shell is fired from one of the pneumatic guns on Presidio Cliff with a velocity of 30,000 feet per minute at an elevation of 47°. If the gun is 400 feet above sea level, compute the time at which the shell will strike the water, the horizontal distance from the bottom of the cliff to the striking point, and the greatest altitude reached.

Maximum Range of a Projectile. If it were not for the resistance of the air a projectile hurled with any velocity would have a maximum range when the angle of elevation is 45°. In Fig. 38,
$$OA = OD \cos \beta,$$
where β is the angle of elevation. But
$$OD = v_p t_a,$$
in which t_a is the time to reach A, or the time that elapses while the body is moving to the highest point and back to the horizontal. And
$$t_a = 2 \frac{v_p \sin \beta}{32.2},$$
whence
$$OA = v_p t_a \cos \beta = \frac{2 v_p^2 \sin \beta \cos \beta}{32.2}.$$
But
$$2 \sin \beta \cos \beta = \sin 2\beta,$$
whence
$$OA = \frac{v_p^2}{32.2} \sin 2\beta.$$

Since $v_p^2 \div 32.2$ is a constant, the range OA is a function of $\sin 2\beta$, and hence OA has its maximum value when $\sin 2\beta$ is greatest. The greatest value that the sine of any angle can have is the sine of 90°, or *one*. Putting $\sin 2\beta = 1$, 2β will then be 90°, or $\beta = 45°$. Therefore, the range OA is greatest when $\beta = 45°$, whatever the velocity of projection.

Section II.

STATICS.

Chapter I.

FORCE. MASS.

In Section I, we learned that the subject of Kinematics is based upon considerations of space and time. In all the remaining branches of mechanics, including the present subject of Statics, we must deal with a new idea; viz., Force,—a clear and correct conception of which is not easily grasped.

In common usage the word "force" has been widely applied and greatly abused. In the preceding chapters we have avoided using it, substituting for it such words as "influence," and "cause," in order that we might begin to use it at a time when we can consider its proper scientific meaning.

If two bar magnets, suspended so as to swing freely, are placed near each other, they will exhibit attraction or repulsion, depending upon the character of adjacent poles. Without stopping to theorize concerning the remote cause of this influence between the two magnets, we can accept the facts that there is an action in which both magnets participate—a mutual action, and that it results in motion or exhibits itself through the motion it produces.

This mutual action between the two given masses is called a **Force**.

An electrified rod of glass, ebonite or other non-conductive material, or even an insulated conductor when similarly electrified by rubbing or otherwise, will attract a stick or rod of metal suspended so as to swing freely. In fact, all the familiar phenomena of electrical attraction and repulsion are instances of an action between two masses quite analogous to the magnetic action above referred to. The remote cause of electrical attraction and repulsion may be very different from the cause of the analogous magnetic phenomena, but, nevertheless, in each case the ultimate effect is an action between the two bodies concerned in the transaction, which moves, or tends to move, both of them. We may have, therefore, a **magnetic force** or an **electrical force**.

If an elastic cord or a spiral spring be elongated, and a ball or other mass be fastened at each end, the cord or spring will contract again as soon as freed, and will even move the two inert masses. The rebound of a rubber ball is due to the same cause. In such cases the moving influence is an action of some sort between the particles of the rubber or of the material in the spring. But it is an action which produces motion, and hence is a force. It differs, however, from an action between two distinct bodies separated by an appreciable distance, as in the case of the bar magnets, because if the rubber band were stretched far enough to break it there would be no perceptible attraction between the two parts. It is some manifestation of cohesion between the individual particles or molecules of the rubber, and the sum total of these molecular actions is what we observe as the total force exerted by the rubber. It may be designated as a **molecular force,** or an elastic force, or by means of any other adjective that will identify it through its origin or through any of its characteristics.

But the commonest exhibition of force is the **action of gravity,** with which we are all familiar in a general way. To our senses this action is evidenced through the phenomena of attraction between large masses, such as the general mass of the earth and objects on its surface, though the accepted theory of gravitation, which will be considered in a later paragraph, gives it a molecular origin. The theories of electrical and magnetic forces would put these also on a similar molecular basis. In all cases, however, whether the size of the body or the quantity of matter under consideration be great or small, the force exhibits itself as an action between two masses.

Definition:—A Force is a mutual action of attraction or repulsion between two masses.

The masses themselves may move under this influence, or the force may be employed to move some other mass, as the stirrups in which the magnets are suspended, or the two balls attached to the rubber cord, or an engine driven by the expansive force of steam.

Resistance. If a person lifts a weight from the ground the force exerted is a muscular contraction (similar to the contraction of a rubber band), which tends to pull the shoulder downward and the hand upward. If the weight is raised we say that the resistance of gravity has been overcome. For the time being we cease, or forget, to think of gravity as an active agent; we regard the muscular force as the action and the force of gravity as a resistance, merely. But reverse the thought: if it were not for the resistance of the hand holding it the weight would move to the earth, and so we think of the hand as resisting the action of gravity. When action is pitted against action, force against force, in this manner, it may suit our convenience to regard either as a resistance to the other.

When a force acts upon a body to bend it, or stretch it, or compress it, or otherwise deform it, there is occasioned in the body a resistance which had no existence until called into play by the action of the force itself. This is a case somewhat different from two entirely independent forces counteracting each other. For instance, a body resting on a table is prevented from falling to the ground because the table resists, or stands against, the action of gravity. The attraction between the weight and the earth compresses the table and crowds its molecules together. The molecules resist this displacement from their normal positions and exert among themselves an expansive force or repulsion, which causes the table to assume its original form and dimensions when the load is removed. Thus it appears that the elasticity of a body may be called into play to furnish a force that will resist an external force. Under such conditions the body is said to be subject to a *Stress*. A stress may be a *Pressure*, or a *Tension*, or a *Shearing Stress*.

A body on a rough surface offers a frictional resistance to any force tending to move it along the surface. This, likewise, is always a passive resistance rather than a counter-action; it can have no existence until called into play by the action of gravity or some other force, and it is merely a recipient of such action.

These are typical illustrations of the phenomena and considerations to be dealt with in Statics. A force being a mutual action of attraction or repulsion between two masses, it always *tends* to produce motion. If it is counteracted, or counterbalanced, in any manner whatsoever, it gives rise to **statical conditions.**

Action and Reaction. The following law, from the time that it was first enunciated by Newton; viz., "To every action there is an equal and opposite reaction," has proved more or less misleading to beginners. Let us investigate some of the causes leading to a misunderstanding of this simple statement. Where there is but one force under consideration—a mutual or dual action between two masses, the application of the idea of "action and reaction" is quite simple. If two bodies, A and B, are connected by an elastic cord, the contraction of the cord pulls equally on A and B, in opposite directions. Or, if the force is not exerted by an elastic cord, but A and B attract each other by gravity or otherwise, then this dual or mutual action may be regarded as a simultaneous operation of two separate reciprocal actions—A upon B, and B upon A, and whichever we choose to mention first, we may call the "action," and the other is its reciprocal action, or its "reaction." To this extent the idea of "action and reaction" simply asserts the dual nature of a force.

It is in cases where one force is balanced by another, such as lifting a weight from the ground, that the complexities and difficulties arise. We have already stated that in lifting the weight the muscular contraction of the arm pulls down on the shoulder and upward on the hand—equally and in opposite directions, like the elastic cord,—giving rise to two reciprocal operations, either of which may be taken as the action and the other as the reaction. This is still a consideration of a single force, and the idea is not yet confused. But, suppose we say that the downward tendency of the weight is balanced by the upward pull of the hand, as already stated in speaking of resistances: can we regard these two opposing influences as a case of "action and reaction?" It is here that the misunderstanding originates,

—when the idea of "action and reaction" is extended to cover this class of cases which we have just termed resistances. It is true that these two opposing influences are equal and opposite, because they balance each other, but they can not produce motion and do not constitute a force according to our definition; on the contrary each is a part or member of two different forces counteracting each other. That is, in the

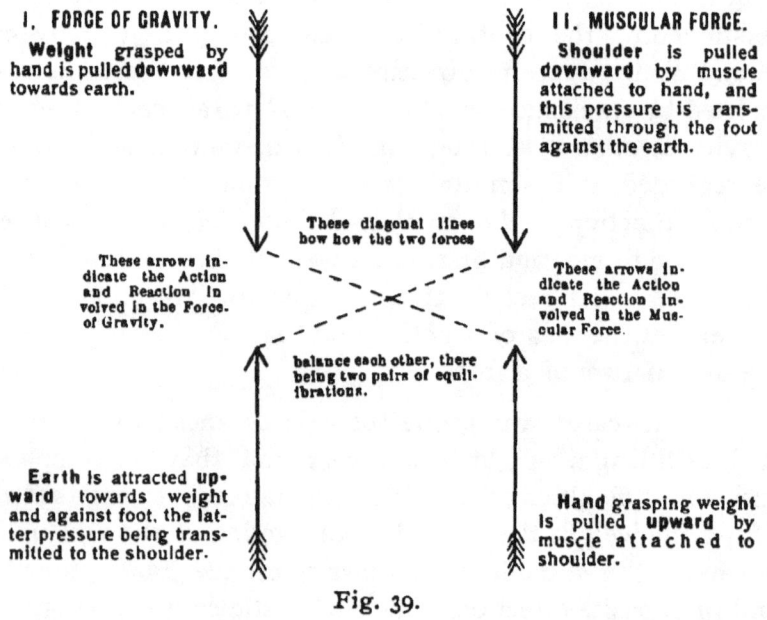

Fig. 39.

case under consideration, we have one of the members of a second force—the upward muscular action on the hand—balancing the downward action of gravity on the weight. But this is only half the transaction; what becomes of the remaining member of each of these two forces? They must also balance each other, if the equilibrium between the two forces is to be complete. That is, if the upward muscular pull on the hand balances the downward pull of gravity on the weight, then the downward muscular pull on the shoulder must balance the upward pull of gravity on the earth. "The actions and reactions of the two forces equilibrate in pairs." Owing to the dual nature of each force it requires a double transaction of resistances in order that the two forces may balance each other. In practice we do not generally take cognizance of more than one of these equilibrations at a time, but the other occurs, nevertheless.

These things are illustrated by diagram in Fig. 39, which shows:

First—That each of the given forces—gravity and muscular contraction—involves an action and its reaction.

Second—That these two forces balance each other completely, giving *two pairs of resistances* or equilibrations.

Similarly, a weight resting on a table is sometimes cited as an instance of action and reaction. The weight pushes downward on the table and the table, it is asserted, "reacts" upon the weight. As a matter of fact the prime active agency in the case is the force of gravity between the weight and the earth, so that if we start with the downward tendency of the weight and call it the "action" then its real reaction is the upward tendency of the earth. As each of these tendencies is overcome by the interposition of the table, we have two pairs of equilibrations—the top of the table *versus* the weight, and the bottom of the legs *versus* the earth. But to say that the table reacts against the weight, and the legs against the earth, is an objectionable use of the idea of action and reaction. It is confounding a *counter*-action with a *re*-action. The action and reaction, or force, between the weight and the earth is not destroyed or even suspended by the intervention of the table, but merely restrained from producing motion.

Composition and Resolution of Forces. A force can be represented, in both magnitude and direction, by a straight line. Component forces may be combined geometrically to determine their resultant, precisely after the manner of the "composition of velocities." Some simple applications of the graphical representation of forces will be employed incidentally, without further explanation, thoughout the subject of Statics.

Law of Gravitation. As already stated, the action of gravity is the most familiar example of force. Electric and magnetic forces are seldom dealt with in the study of theoretical mechanics, the problems and illustrations usually referring to weights, and to pressures, tensions, etc., due to weights. The familiar facts of gravity are these:

That the weight of a body depends upon the amount of material in it; the weight is directly proportional to the mass.

That forces of all kinds,—electric, magnetic, muscular, etc.,—are usually balanced by weights, and hence are measured in gravitational units, such as pounds, ounces, grams, kilograms, etc.

That weight acts vertically downward towards a point at or near the center of the earth.

That the weight of a given body depends somewhat upon its distance from the center of the earth, being greater at the poles than at the equator, and greater near the sea level than at high altitudes.

That the rotation of the earth diminishes the weight of bodies, with greatest effect at the equator and reducing to zero at the poles.

While these general considerations may be clear in a general way, yet a proper appreciation of the Law of Gravitation requires considerable practice in the use of the so-called mathematical **law of inverse squares**, as applied to problems of celestial mechanics, such as planetary attractions.

FORCE.   65

Mathematical Expression of the Law of Gravitation.
The Law of Gravitation asserts that between every pair
of particles in the universe there is a mutual attraction,
which varies (1) directly as the product of the masses of the
particles, and (2) inversely as the square of the distance
between them.

(1) If A represents the force of attraction between two
particles of matter, m and m', then (disregarding distance
for the present, and using the sign of variation to avoid the
necessity of selecting units of force and mass), according to
the first part of the law,
$$A \propto m \times m'. \qquad *$$

The particles m and m' may be alike, or they may be as
different as we please; they may even be magnetized or
electrified, and thus have a second attraction (or repulsion)
entirely independent of gravity. Their gravitational
attraction may be represented by diagram, thus:

Fig. 40.

If m be replaced by two particles of the same kind, or
by a single particle of twice the mass, the attraction will be
doubled, thus:

Fig. 41.

If both m and m' be doubled the attraction becomes four
times as great, (the distance, of course, being assumed to
remain unchanged), thus:

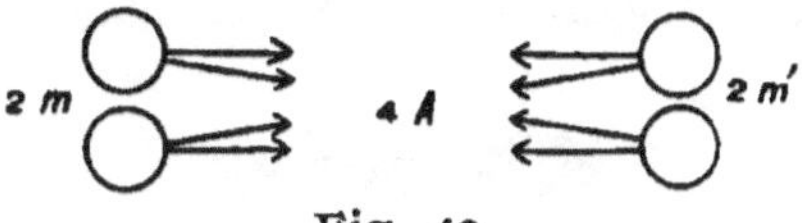

Fig. 42.

If m be increased to $3m$ and m' to $5m'$, the attraction
will become $15A$; etc.

---
* The symbol $\propto$ means "varies with."

(2) If the two original particles be not changed in mass, but are moved to different distances from each other, then as this distance d is changed

$$A \propto \frac{1}{d^2}.$$

If the attraction is A at distance d (Fig. 43),

Fig. 43.

at distance $2d$ (Fig. 44) it will be $\dfrac{A}{4}$.

Fig. 44.

At distance $3d$ it will be $\dfrac{A}{9}$; etc. If d be diminished to $\tfrac{1}{2}d$, it will be $4A$.

If the masses and distance both change simultaneously, what is the result? For instance, if m be doubled and m' tripled, and the distance doubled at the same time, the increased masses will multiply the attraction 6 times, but the increased distance will diminish the result to one-fourth of what it would have been if the distance had remained unchanged. The result will be an attraction $\tfrac{6}{4}$ as great as the original A between m and m' at distance d. That is,

$$A \propto \frac{m\,m'}{d^2}$$

The attraction between two large spherical masses of uniform density is the same in effect as if the entire material in each mass were condensed into its central particle. For instance, in computing the attraction between objects and the earth, we consider the distance to the center of the earth and not to the surface.

Examples:

1. *A body, or aggregate of particles, has a mass of 8 units, and a second body has a mass of 15 units. At a distance of 12 units from each other, these bodies exhibit a gravitational attraction equal to 50 grams. Express the attraction between each of the following pairs of bodies at the distances specified:*

	Mass of each of given bodies.		Distance between bodies.
	I	II	
(i)	5 units	6 units	24 units
(ii)	5 "	6 "	48 "
(iii)	2 "	10 "	6 "
(iv)	30 "	3 "	2 "
(v)	40 "	7 "	8 "
(vi)	25 "	13 "	18 "

2. *The attraction between two masses, m and 2m, at distance d from each other, is equal to A. Another pair, of magnitudes 2m and 3m, are placed at a distance 3d from each other; what is the attraction between them, expressed in terms of A?*

3. *A body on the earth's surface weighs one pound.*
    (a) *What will it weigh 4000 miles above the earth's surface?*
    (b) *What will it weigh 8000 miles above the earth's surface?*
    (c) *What will it weigh 6000 miles above the earth's surface?*

It is assumed that the body is always weighed on a spring balance graduated at the earth's surface. Why?

4. *The earth is said to be slowly cooling and contracting. How does this affect the weights of bodies on the earth's surface? Why? What would be the effect if the earth were expanding? Why?*

5. *Assuming that the mass of the Moon is $\frac{1}{81}$ that of the earth, is there a point between them at which a body would have no weight? If so, locate it.*

6. *A spherical body weighs 10 pounds at a given place on the earth's surface. Find the weight at the same place of a second sphere of twice the radius and three times as dense as the first body.*

7. *A given body weighs one pound on the earth's surface.*

(*a*) *What will it weigh on another planet of the same density and whose mass is 27 times that of the earth?*

(*b*) *What will it weigh on still another planet of the same density, but the mass of which is $\frac{27}{64}$ that of the earth?*

8. *A spherical body weighs 100 pounds on the earth's surface. Find the weight of a second body of twice the radius and twice as dense, situated on the surface of a planet of three times the earth's radius, but only half as dense as the earth.*

Weight of a Body Beneath the Earth's Surface. It has been shown that a body carried above the earth's surface loses weight inversely as the square of the distance from the earth's center, but we should not infer from this that a body would gain weight if it could be carried in the opposite direction, towards the earth's center. On the contrary, it is obvious, that if the earth were a sphere of uniform density, a particle at the center would be attracted equally in all directions, and furthermore, it can easily be demonstrated that a body in transit from the surface to the center would lose its weight uniformly with the distance. The weight of such a body would vary *directly as the distance* from the center, and *not inversely as the square of the distance*. While this demonstration involves practically nothing more than the law of gravitation, it gives rise to the application of the law of inverse squares to the following important proposition:

A body placed at any point within a hollow spherical shell of uniform density is at perfect equilibrium as regards the attraction of the shell.

In Fig. 45, let P be a particle at any point within the shell shown in section in the diagram. Let PB and PC represent elements of a right circular cone. If the surface of this cone were continued, it would cut out of the

FORCE. 69

spherical shell a concave disc represented in section as AB. BP and CP continued will form elements of a second and similar right circular cone, which would cut from the shell a second disc of section A_1B_1. (We shall speak of these sections in place of the area which they represent).

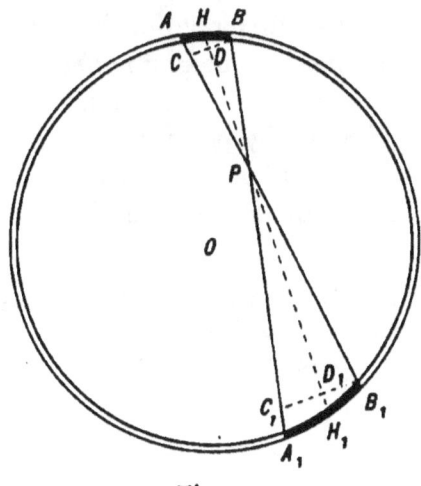

Fig. 45.

By Prop. VI, Book IX of Euclid, (Wells' Geometry, p. 355), the areas of the bases BC and B_1C_1 of these cones are to each other as $\overline{PD}^2$ is to $\overline{PD_1}^2$. As the vertex angle of the cone diminishes, the bases BC and B_1C_1 approach the areas AB and A_1B_1, respectively, as limits. Therefore, for any small area, as the limiting area AB, there is a corresponding area A_1B_1 so situated that

$$\frac{\text{area } AB}{\text{area } A_1B_1} = \frac{\overline{PH}^2}{\overline{PH_1}^2}.$$

Under these circumstances it is obvious that the attraction between the particle at P and the mass of AB is just equal and opposite to the attraction between the

70   THEORETICAL MECHANICS.

particle at P and the mass of A_1B_1. For instance, if the mass included within the area A_1B_1 is four times as great as the mass of AB, it can be so only because the *square* of the distance PH_1 is also four times the *square* of PH; the predominance of mass A_1B_1 is exactly neutralized by its greater distance (squared).

If it were not that this relation of dimensions of the spherical shell happens to coincide in effect with the law of inverse squares as applied to attractions, a particle would not be at equilibrium at any point within the shell, except at the very center; no other law of attraction would give the general condition of equilibrium for all points within the spherical shell.

An electrified pith ball remains at rest at any point within an electrified spherical shell; showing that the law of inverse squares is also applicable to attractions and repulsions between electrified bodies.

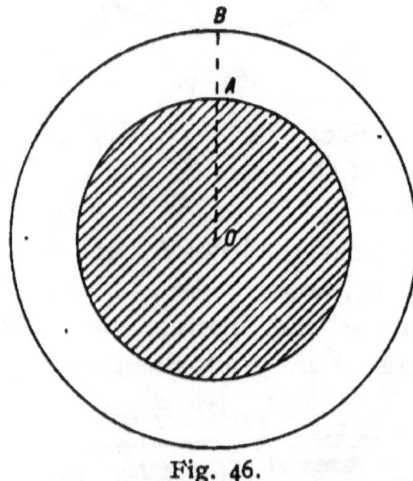

Fig. 46.

Now, if a body is carried towards the center of the earth, at any point *en route*, as at A in Fig. 46, the entire spherical shell of thickness AB can be regarded as exerting no attraction whatever upon the body. The resultant attraction upon the body is the same as if the earth were a planet of radius OA. If $OA = \frac{OB}{2}$ and the density of the earth were uniform at all depths, then the mass of the shaded portion would be ⅛ of the total mass of the earth. Hence, while the body at A would be only 2000 miles, (or ½ as far as B) from the earth's center, on the other hand it would be attracted by only ⅛ of the mass, and the apparent weight would be $2^2 \times$ ⅛, or ½ of its weight at the surface.

That is, the portion of the earth which acts as an attracting mass varies directly as $\overline{OA}^3$, and the effect of distance inversely as $\overline{OA}^2$; hence the weight varies as $\dfrac{\overline{OA}^3}{\overline{OA}^2}$, or directly as OA.

In example 6, p. 67, in which a body was weighed on different planets, the weight of the body varied directly as the radius of the planet.

This demonstration assumes that the earth is of uniform density at all depths, but the fact is that the density increases towards the center. Observations indicate that bodies carried downward exhibit an actual increase of weight instead of a decrease, down to a certain depth, beyond which the weight begins to diminish. It appears, therefore, that a body really has its maximum weight, not at the surface, but a short distance below.

Mass. As stated in the Introduction, p. 2, we have not considered a rigid definition of mass, though we have used the related ideas of weight, density and specific gravity. Our use of this word, therefore, has been more or less tentative. Our assumption has been that every body has in it a definite amount of matter, and we judge this amount by the weight of the body. To judge the mass of a body from its weight is safe enough where we experience no greater changes in the intensity of gravity than we encounter from place to place on the earth's surface. But if we could be suddenly transported to the surface of the moon with the given body its apparent weightiness would be greatly diminished, though its mass has remained unchanged. In fact, as we learned from the law of gravitation, the mass of a piece of lead even would have no weight at all if it were not for the existence of some other attracting mass, such as the earth. But we cannot conceive of the lead losing its mass; the very existence of the body is made known to our senses through its mass, or the amount of material existing in it. For this reason mass is said to be one of the *essential* properties of matter,—essential to our conception of its existence.

Masses are designated as pounds, ounces, grams, kilograms, etc.,—the same as weights and forces. This identity of names as applied to different things arises, of course, from relations between those things, but it is confusing to a beginner who has not yet learned those relations. According to the law of gravitation the weight of a body at any given place, (that is, its attraction towards the earth), is proportional to its mass. Expressed mathematically, *the weight* $\propto$ *the mass*. If we choose, we may take a lump of any kind and call it a unit mass, and then we may take as the unit weight, the weight of this unit mass, or of any fraction or multiple of it, at any standard place as regards latitude and altitude. A lump of platinum* preserved at the Archives of Paris as a standard

---

*This is supposed to contain the same mass as a litre of water at its temperature of maximum density, 3°.9 C., but it has been found that a litre of pure water at 3°.9 C. really weighs 1.000013 Kg.

mass of one kilogram also *weighs* one kilogram *at Paris ;* it weighs more than a kilogram at the North Pole, and less than a kilogram at the Equator.

The standard pound mass is a piece of platinum preserved in the office of the Exchequer at London ; the weight of this mass *at London* is the standard pound weight.

Density. The density of a given substance is the **mass of a unit volume** of that substance,—expressed in grams per c. c., pounds per cu. foot, etc. Notice that it is the mass of a unit volume, and not the weight. The weight of the body may be different at different places ; it may even be zero, as at the center of the earth, in which event if the density were regarded as the *weight* of a unit volume, we would have to imagine the body to exist without density.

Heavy and Light Bodies Fall at the Same Rate. In discussing the acceleration of gravity we assumed that all bodies fall at the same rate. This is true if we disregard the frictional resistance due to the presence of the air. Obviously, a light feathery body or a piece of paper has to displace and drag itself along through a quantity of air that is very large in proportion to the weight of the body. In a vacuum a feather falls as fast as a piece of metal.

A little thought will show that it would not be consistent with our common experience for a heavy body to fall faster than one half as heavy. If two bricks are dropped, one from each hand, they will reach the ground in the same time ; if the two had been glued together, there would be nothing to make the double mass fall any faster than the single bricks. In dynamics this would be explained by saying that while the double mass has twice the attraction downward, it also has twice the mass to be moved by that attraction.

CHAPTER II.

WORK. POWER. ENERGY.

We know that to lift a weight or drag a vehicle, or to overcome a resistance of any kind, requires the application of a force. If we consider this applied force in relation to the distance through which the body moves under its influence, the product of these two values—the force and the distance—introduces one of the most important conceptions in mechanics. In lifting a pound weight one foot, it is said that one **Foot-Pound** of **Work** is done. To lift a five pound weight from the floor to the top of a table three feet high, requires the performance of 15 foot-pounds of work. If a vehicle requires a pull of 130 pounds to drag it along a horizontal surface the work done for every mile that it is moved along such a surface is 130×5280 foot-pounds.

Of course, there may be more than one unit of work, as a mile-pound, or a mile-ton; a kilogram-meter, or a gram-centimeter,—according to convenience for large and small measurements, and to meet other needs. For practical purposes, however, it is seldom that any unit of work is employed other than the foot-pound for the English system, and the kilogram-meter for the metric system.

The work is said to be done *by* the applied force and *against* the resistance.

The time consumed is not a factor in determining the amount of work done in overcoming a resistance through a certain distance. The work done in carrying a thousand bricks to the top of a given building is the same whether it is accomplished in a day or a week. Work would be done faster in one case than in the other, but the total number of foot-pounds is the same.

In the sense of our definition, no work is done when a weight is held at rest in the hand, or when it is moved horizontally. In the first instance, it is obvious that the body is moved through no distance, so that the force multiplied by the distance is *zero*. In the second case, the body is moved, but it is neither raised nor lowered. The amount of work done is determined by the distance through which the resistance yields to the applied force. With the weight held in the hand the muscular action and the action of gravity are both vertical, and as there is no vertical motion there is no work done. The work done in carrying a weight up a flight of stairs is the same as if the body were lifted vertically from the lower floor to the next. Work done against gravity depends only upon difference of level.

To make this point sufficiently clear, it seems almost necessary to anticipate a principle of Dynamics. It was stated on p. 50 that, if it were not for the action of gravity, a free and unobstructed body hurled in any direction would move on forever in a straight line with uniform velocity; that is, no force would be required to keep it moving, and hence no work would be done as it moved mile after mile through space. Even with gravity acting, if it were not for friction, it would not require work to keep a vehicle moving on a horizontal plane, once it were started. On a *perfectly smooth* horizontal plane (if such could be realized), the slightest force would start any mass, howsoever large; as long as this force continues to act, the body would keep on moving faster and faster; when the desired speed is attained the force could be withdrawn, or cease to act, and the body would continue to move on indefinitely with this velocity. In other words, when a weight is moved on a horizontal surface it is the resistance of friction that must be overcome and against which work is done; the action of gravity between the weight and the earth is not overcome, and hence no work is done against it. Indirectly the weight counts in this way, that the friction caused by such a body would be directly proportional to the weight, and for that reason more work would be done in moving a heavy body than in moving a lighter one through equal distances on the same horizontal plane.

Examples:

1. *A person weighing 150 pounds walks up a flight of stairs between two floors 14 feet apart. What work is done? What if a vertical ladder had been used? What if he had climbed up on a rope?*

2. *A horse drags a plow a half mile against a resistance of 200 pounds, as indicated on a dynamometer. How much work is done?*

3. *A person walking against the wind has to overcome a resistance of, say, 4 pounds per square foot. If the surface meeting this resistance is 5 square feet, what work does he do for every mile walked?*

4. *A gallon contains 231 cubic inches. A cubic foot of water weighs 62.5 pounds. What work will be required to pump 1000 gallons of water through a vertical height of 50 feet?*

5. *How many foot-pounds of work will be done in raising a five-gallon can of alcohol from the floor to the top of a table 33 inches high? (Specific gravity of alcohol = 0.8).*

6. *A person weighing 155 pounds rides up a hill on a bicycle weighing 27 pounds. If the hill rises $10°$ from the horizontal, what work is done for every 100 feet ridden?*

7. *In a steam engine of 9.5 inch diameter of piston and 12 inch stroke, how much work is done during each complete stroke if the effective steam pressure in the cylinder averages 30 pounds per square inch? If the engine is running 200 revolutions per minute, how many foot-pounds of work are done in a second?*

8. *A kilogram-meter is equivalent to how many foot-pounds?*

Power. The horse-power of an engine is determined by the *rate* at which it can do work. The idea of Power simply modifies the idea of Work by introducing a time element. To raise a ton of granite to the top of a building 75 feet high requires 150,000 foot-pounds of work, without regard to the time consumed. But the engine that can accomplish this amount of work in one minute has twice the power of an engine that would require two minutes for the same work.

A natural or logical unit of Power would be the ability to do one foot-pound of work per second, or some similar value employing any convenient units of force, distance and time. It is the practice, however, to use an entirely arbitrary unit—**the Horse-Power.** An engine or other agency works at the rate of one H. P. if it performs 550 foot-pounds per second, or 33,000 foot-pounds per minute.

The horse-power of an engine under a given pressure of steam could be readily computed from its dimensions and speed, if the full steam pressure acted throughout the entire length of stroke, or even if the average effective pressure were known. For example, assume the following conditions: Diameter of piston 9.5 inches; length of stroke, 12 inches; number of revolutions, 200 per minute; boiler pressure, 75 pounds per square inch. If the full steam pressure acted throughout the entire stroke, the work done during each double stroke, or revolution, would be

$$2 \times 75 \times \frac{\pi \times \overline{9.5}^2}{4} \times \frac{12}{12} \text{ foot-pounds.}^*$$

The work done per minute would be

$$2 \times 75 \times \frac{\pi \times \overline{9.5}^2}{4} \times \frac{12}{12} \times 200 \text{ foot-pounds,}$$

and the horse-power supplied to the piston from the energy of the steam would be

$$2 \times 75 \times \frac{\pi \times \overline{9.5}^2}{4} \times \frac{12}{12} \times 200 \times \frac{1}{33,000}, \text{ or } 64.4 \text{ H. P.}$$

---

*Assume $\pi = 3.1416$.

As a matter of fact the mean effective pressure on the piston is considerably less than the pressure of the steam as it enters the cylinder. The inlet valve of an engine is adjusted in such a manner that the steam supply is shut off when the piston has completed only a fraction of its stroke—a third, or a fourth, or at whatever point may be necessary for the best economy under the given conditions of steam pressure and load. After the steam is cut off, the remaining part of the stroke must be completed by the expansion of the steam previously supplied, and of course, the pressure of this steam diminishes as its volume (the space it occupies in the cylinder) increases.

It is the function of the governor to vary the supply of steam as needed to maintain a constant number of revolutions per minute. There are two types of such governors. If the valve is fixed so that it always cuts off the supply at the same fractional part of a stroke, then the governor regulates the steam supply by varying the size of an opening through which the steam is made to pass *en route* to the cylinder. The common Watt governor is of this type. In an automatic cut-off engine the governor controls the valves themselves, varying the point of cut-off.

The valves of an engine are also adjusted to control the outlet of steam, or the exhaust. To accomplish easy running, enough steam should be confined in the idle end of the cylinder at each stroke to furnish a cushion and thus prevent the sudden stopping and consequent jarring at the end of the stroke. To compress this steam, of course, neutralizes some of the pressure in the active end of the cylinder, so that the mean effective pressure is still further reduced from this cause.

In the example selected for illustration, on the preceding page, it was found that the steam would convey to the engine 64.4 H. P., if the total pressure of 75 pounds per square inch were to act throughout the entire stroke of the piston. Now, on p. 47, it was stated that this same engine, under exactly the same conditions, indicated 24.77 H. P., as determined from an indicator card. This shows that the mean effective pressure on the piston was less than the boiler pressure in the ratio of $\frac{24.77}{64.4}$. If P_m is the mean effective pressure, then

$$P_m \times \frac{\pi \times \overline{9.5}^2}{4} \times \frac{12}{12} \times 200 \times \frac{1}{33,000} = 24.77,$$

or, $P_m = 28.84$ pounds per square inch, which will be found to be $\frac{24.77}{64.4}$ of 75.

Some of the energy transferred to the piston from the steam is wasted by friction in the engine itself. The rate at which energy is supplied to the piston is called the Indicated Horse-Power, as contrasted with the Actual Horse-Power available from the engine.

The ratio of the Actual Horse-Power to the Indicated Horse-Power, expressed as a percentage, is the efficiency of the engine.

Electrical power is measured in **Watts**, this unit being the rate at which energy is conveyed by a current of one ampere intensity under a potential of one volt. One H. P. in mechanical measure is equivalent to 746 Watts.

Examples:

 1. *How many foot-pounds of work can be done in 8 hours by a 50 H. P. engine?*

 2. *How many tons (2000 pounds) of coal per hour can be raised from the ground to a bin 60 feet above by a 10 H. P. engine?*

 3. *What H. P. will be required to pump water to a height of 120 feet at the rate of 1000 gallons per minute?*

Assume 231 cubic inches to a gallon and 62.5 pounds as the weight of a cubic foot of water.

 4. *If a horse can perform 550 foot-pounds of work per second, at what rate in miles per hour can he drag a plow against a resistance of 225 pounds?*

 5. *A person carries a sign measuring 3 feet by 4 feet, against the wind, the resistance being 6 pounds per square foot. If he walks 2 miles per hour, at what H. P. is he doing work?*

 6. *A person weighing 150 pounds runs up a flight of stairs in 5 seconds. If the stairs are 27 feet long and rise at an angle of 34°, the person is doing work at the rate of what H. P.?*

7. How many gallons of water will a 40 H. P. engine pump in an hour from a mine 500 feet deep?

8. A house on rollers is moved by means of pulleys and a windlass. If the resistance to rolling is 20 tons, at what rate can it be moved by a single horse working at the rate of one H. P.?

9. What must be the H. P. of an engine if it is to be used for running a 110 volt dynamo that generates a current of 50 amperes?

10. What is the H. P. of an electric motor driven by a current of 75 amperes under potential of 125 volts?

11. Compute the H. P. developed by an engine under the following conditions: Diameter of piston, 4 inches; length of stroke, 5 inches; mean effective pressure, 42 pounds per square inch; number of revolutions, 275 per minute.

12. An engine has the following dimensions: Diameter of piston, 12 inches; length of stroke, 18 inches. At 150 revolutions per minute the Indicated H. P. is 35.2. Find the mean effective pressure.

Energy. It has been stated that work is regarded as being done *by* the applied force and *against* the resistance. When a person lifts a weight the muscular action is clearly the applied force and work is done against the resistance of gravity. But suppose that in this elevated position the weight is attached to a clock, or other piece of mechanism; by the action of gravity the weight descends and does work against the resistance of the machine driven by it. If it is a 10-pound weight and was lifted 2 feet above the floor, the work done in raising it was 20 foot-pounds; this is exactly equal to the amount of work that can be done on the machinery by the weight as it returns to the floor. The work done in lifting the weight was not wasted; it was simply invested in the elevated mass as available **Energy**, ready to be given back in full by performing 20 foot-pounds of work. A body that is in any way endowed with the ability to do work is said to possess energy. If it can do 100 foot-pounds of work it possesses 100 foot-pounds of energy.

Potential Energy. Hydraulic elevators are operated by storing water, either in an elevated tank or in a closed tank under pressure. Many machines are driven by compressed air, the work done in compressing the air being given back as the molecules return to their normal distances from each other. The energy stored in a clock-spring by winding is of a sort much the same as the energy of compressed air; in all such cases advantage is taken of the elasticity of the material, which offers a resistance to any force tending to deform it and thus stores up any work done upon it. These are all examples of Potential Energy, or as it is sometimes called **Energy of Position.**

Two bodies that attract each other, such as the weight and the earth, are endowed with this energy only by the act of separating them. The energy invested in the bent spring or the compressed air, or in any other elastic body, is due to the displacement of the particles from their normal positions. It is the tendency of the disturbed bodies, or the disturbed particles of the distorted body, to return to their normal positions that gives them the ability to do work.

The adjective "Potential" means "possible" and is used to signify that the exhibition of this energy is contingent upon the condition that the displaced body or particles be thus allowed to return to their normal position or positions. For instance, if the elevated weight is placed on a shelf or a table-top it can do no work in that position, but must be allowed to fall to the ground, if its energy is to become evident. The driving weight of a clock is held by a catch and performs its work only as it is released by the escapement. The energy of a bent spring and of compressed air is potential because of the same contingency.

The potential energy of a body, as ordinarily understood, is not an absolute quantity. The potential energy of an elevated weight, for example, is measured relatively to the floor, or to the ground, or some other plane taken as a standard.

Kinetic Energy. As contrasted with Potential Energy or Energy of Position, a body may have Kinetic Energy **due to its velocity.** For illustration, a jet of water can be made to drive a water-wheel and thus perform work; the work done by a wind-mill in pumping water is readily traced to the energy of the air current; it is the kinetic energy of the carpenter's hammer that does the work of driving a nail.

Energy of motion differs from energy of position, also, in this respect that the former exists free from all contingency. For that reason Kinetic Energy is sometimes called **Actual Energy,** in distinction from Potential (or possible) Energy.

The kinetic energy of a body is the same, no matter what the direction of motion. It could be *measured* experimentally by fastening the body in such a way as to compel it to overcome a known resistance or let it strike some properly arranged obstacle, and then observing how far the resistance yields before the body comes to rest. The simplest way to *compute* the kinetic energy of a body from its mass and velocity is to find out how far it would rise *if it were moving vertically upward.*

For example, suppose a given body to have a velocity of 96.6 feet per second, and we wish to find its kinetic energy.

Now we know from the principles of acceleration that this body, if it were moving vertically upward and started from the ground with this velocity, would rise to a height of 144.9 feet. In other words, by virtue of its velocity, it would be able to lift itself to this height against the action of gravity. Its weight multiplied by this height is, therefore, the amount of work it can do, or its energy.

As a further example, if a bicyclist wished to know his kinetic energy at a certain speed, he could start up a hill of known pitch with this velocity and, removing his feet from the pedals, note how far he is carried up the hill. His weight multiplied by the *vertical* height to which he ascends will represent his kinetic energy at the foot of the hill.

In formula 8, p. 44, it was proved that the distance a body would travel while gaining (or losing) a given velocity varies as the square of this velocity. Hence the kinetic energy of a body depends upon the square of its velocity. One body traveling 5 times as fast as another body of the same weight would have 25 times as much kinetic energy.

In Heat, Sound, Light and Electricity, we find examples of **Molecular Energy,** as distinguished from **Mechanical Energy,** which heretofore we have drawn upon exclusively for purposes of illustration. "The Heat possessed by a body is explained as being the Energy possessed by it in virtue of the motion of its particles. Just as a swarm of insects may remain nearly at the same spot while each individual insect is energetically bustling about, so a warm body is conceived as an aggregation of particles which are in active motion while the mass as a whole has no motion."* Such a body has invisible Molecular Kinetic Energy. Moving *en masse*, with visible motion, its energy would be

---

*Daniell's Text Book of the Principles of Physics, p. 48.

called Mechanical. Obviously, it may have both at the same time. Or, if raised to an elevated position, its mechanical energy would be potential,—its molecular of course, remaining unchanged.

The **Chemical Energy** of gunpowder is a familiar example of *Potential* Molecular Energy,—though perhaps it would better be called Atomic rather than Molecular. The different constituents of gunpowder—sulphur, saltpetre and charcoal—have a chemical affinity for each other, by virtue of which they tend to come together and form new compounds, just as a weight tends to fall to the earth. In the gunpowder the constituents, while intimately mixed together, are still separated from each other chemically, and hence they have potential energy. Once the impulse is given, like releasing the elevated weight from the shelf, these chemical constituents rush together, atom clashing against atom, forming new gaseous compounds, and generating an enormous amount of heat. Some of the original chemical energy (molecular potential) is thus converted into Heat (molecular kinetic).

Any form of molecular energy is measurable in mechanical units.

The Watt, as already stated, is equivalent to $\frac{1}{746}$ H. P. This is really a conversion of Electrical into mechanical *power*, and *not* a conversion of *energy*. The mechanical equivalent of an electrical unit of energy could easily be derived, but is not needed in dealing with *currents* of electricity, because a current always involves a time element, which we also found to be involved in power but not in work.

The unit of Heat is the amount of heat necessary to raise a unit weight of water one degree in temperature. Taking a pound as the unit of weight and using the

Fahrenheit thermometric scale, the heat necessary to raise a pound of water 1° F. is equivalent to 778.5 foot-pounds of work.

Transference and Transformations of Energy. Any one of the different kinds of energy can be transformed, directly or indirectly, into any other kind. Mechanical energy is readily changed from the kinetic form to the potential, and *vice versa*. Mechanical energy of either form can be converted into heat, sound, light or electricity. And any one of these various molecular forms can be transformed into any other, or into mechanical energy. Chemical energy is likewise convertible, especially into heat and electricity.

Energy can also be transferred from one body to another and thus transmitted from place to place. Whenever work is done there is a transformation or a transference of energy, or both. In fact, at this point our **definition of work** might well be revised in accordance with this larger view.

But with all these changes of energy from body to body and from one form to another, there is no actual gain or loss of energy. It is always a case of adding and subtracting the same quantity. The work done *by* one body exerting a force on a second body is just equal to the work done *upon* this resisting body, or in other words the first body yields up to the second body an amount of energy equal to the number of foot-pounds of work done. So far as we know the total energy of the universe is an unchanging quantity—a definite number of foot-pounds. This hypothesis is known as the **Conservation of Energy,** and is accepted as the most fundamental principle of physics.

A simple illustration is in the conversion of mechanical energy from the kinetic form to the potential, and *vice versa*. Suppose that a ten pound weight is projected vertically upward with a velocity of 96.6 feet per second. At the instant of leaving the ground it has 10 × 144.9 or 1449 foot-pounds of *kinetic* energy, because by virtue of its velocity of projection it is capable of rising to a height of 144.9 feet against a gravitational resistance of 10 pounds. But when it has reached the highest point, ready to start downward—what then? In this position it has 1449 foot-pounds of *potential* energy. And if at that instant the weight could be caught on a hook or a shelf, it could then be attached to a clock or other mechanism and its potential energy used for any purpose desired,—in which event the body would return to the ground leisurely and perhaps with a velocity hardly perceptible. But if it is not thus restrained, but is allowed to fall freely from this height of 144.9 feet, it will fall in the usual manner—at an accelerated rate. On the downward trip it will gain velocity as fast as it lost it on the upward trip, and when it reaches the ground it will have the same kinetic energy that it started with. Instantly it strikes the ground, however, this mechanical energy of motion is converted into sound, heat, and perhaps light and electricity. It follows, therefore, that when a body is projected upward its kinetic energy is *gradually* converted into potential energy, and back again to the kinetic form as it descends. At the summit it has only potential energy. For every foot that it rises, if its weight is 10 pounds, it gains 10 foot-pounds of potential energy and loses an equal amount of kinetic energy. Half way up half its energy is potential and half kinetic. The total energy, however, remains the same.

The generation of an electric current affords another good illustration of transformation. The current itself is, of course, an example of transmission or transference of energy. If the current is generated by a battery—a voltaic current—it is a case of direct conversion of chemical into electrical energy. If it is furnished by a dynamo—a magnetic-electric current—its energy can be traced back step by step, first to the mechanical energy of the machinery, and thence to the energy of the steam generated by the heat of the furnace. The heat itself originated in chemical energy of the coal and oxygen uniting with each other to form carbonic acid and the other gases. We might go a step further and say that the energy of the coal and oxygen came from the rays of the Sun which in some past age decomposed a lot of carbonic acid gas, giving the carbon to make the wood of a tree and liberating the oxygen to form part of the atmosphere.

ENERGY. 87

This dynamo might have been driven by a water wheel, power being furnished by a jet of water. And perhaps the water was first stored for some length of time in a reservoir. Looking still further we find that it was raised to the reservoir from the ocean at the expense of energy of the Sun.

There is hardly an instance in which the power used by man could not be traced to the Sun. The energy radiated from the Sun itself may be due, all or in part, to chemical action now going on there or to radiation from a molten mass, or to still other causes equally probable and based upon even more recent theories.

Examples:

1. *What is the potential energy of a body weighing 100 kg. at an elevation of 22 meters?*

2. *A body weighing 5 pounds is dropped from a height of 1000 feet. What kind of energy and how much did it have when it started? At the end of one second what change will have taken place? At the end of two seconds? At the end of three seconds? Just as it reaches the ground? After it has struck the ground?*

3. *A body is projected vertically upward with a velocity of 322 feet per second. How much energy has it and of what kind? What change will have taken place at the end of one second? At the end of two seconds? At the end of three seconds? At the end of ten seconds?*

4. *What transformations of energy take place during the oscillations of a pendulum?*

5. *A bicyclist weighing 155 pounds mounted on a 27-pound wheel rides at a velocity of 20 miles an hour on a horizontal plane.*

    (*a*) *What is the combined kinetic energy of the man and wheel?*

    (*b*) *If he comes to a slope rising at an angle of 12° and removes his feet from the pedals, how far will he progress up the hill?*

6. *The same person riding at the same rate comes to a hill of unknown pitch, and finds that his velocity is sufficient to carry him 75 feet up the slope. Find the pitch of the hill from the horizontal.*

7. *In the Yosemite Water-Fall the total drop is about 3000 feet.*

  (a) *What transformations of energy occur?*

  (b) *When the water strikes at the foot of the Falls its mechanical kinetic energy is converted mainly into heat. If there were no loss from friction, etc., during the drop, and if all the energy were converted into heat at the instant of striking, how much would the temperature of the water be raised?*

8. *A body has a velocity of 100 feet per second and weighs 21 pounds. What kinetic energy has it?*

9. *A body weighing 50 pounds and having a velocity of 40 feet per second moves along a horizontal plane against a frictional resistance of 2 pounds.*

  (a) *How far will it travel before coming to rest?*

  (b) *How long before it will come to rest?*

HINT: First compute the energy of the body by finding out how far it would rise against its own weight *if it were moving vertically upward.*

10. *A freight car weighing 30 tons is moving on a horizontal track at the rate of 40 miles per hour. If the total resistance occasioned by the brakes and friction on the tracks is 2 tons, how far will the car move before it is brought to rest?*

11. *What transformations of energy take place when a firecracker explodes? When a steam-whistle blows? When a gong is sounded?*

12. *An electric current is used to drive a motor attached to a lathe. What transformation of energy takes place? What if the current had been used for an incandescent lamp? For electro-plating? For ringing an electric bell? For electric welding?*

Graphical Representation of Work. Using two rectangular axes in the manner explained on pp. 46 and 47, Work can be represented as an area—a rectangle, of which one side or dimension stands for either the applied force or the resistance and the other dimension stands for the distance through which the resistance is displaced. For example, to represent the work done in raising a weight of 600 pounds to a height of 50 feet, let us assume a vertical scale of 1 inch = 200 pounds and a horizontal scale of 1 inch = 10 feet. From the origin O, Fig. 47, measure off on the Y-axis the length $OA = 3$ inches (to represent the weight, 600 pounds), and on the X-axis measure off the distance $OB = 5$ inches (to represent the displacement of 50 feet).* Since OA represents a force and OB a

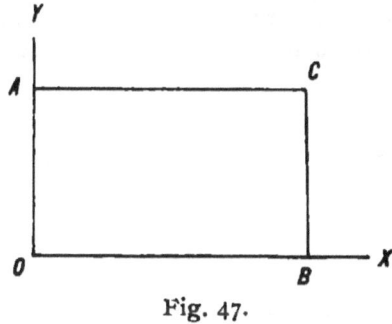

Fig. 47.

displacement caused by this force, it follows that the work done is represented in the diagram by the area of the rectangle $OBAC$, or what is the same thing $OA \times OB$.

If the diagram is analyzed still further it will be seen that this conclusion is an entirely consistent one. Since the weight remains constant throughout the displacement, a perpendicular erected at B or at any point between O and B, and equal to OA, will represent the magnitude of the force for that particular point in the displacement. The *locus* of the extremities of such perpendiculars will be a line through point A parallel to OB.

If the force had been variable the *locus* AC would be an irregular line as yy_1 in Fig. 36, p. 49. The work done, however, would still be the area included between this line and the two axes, or Oxy_1y in Fig. 36. In fact, this general method of interpreting an area is the

---

*The printed diagram is reduced to one quarter of this scale.

same as was explained on pp. 48 and 49 in connection with accelerated motion, except, of course, that the co-ordinates in that case were made to represent entirely different quantities from those we are now dealing with in connection with Work.

Between these two extreme cases of a uniform force and an irregularly changeable one, there are circumstances under which the force changes uniformly, and for which the diagram yields a **triangular area** to represent the work done. This is always the case when a force is applied so as to gradually elongate a wire or rod, or a spiral spring, and in all other cases **where a stress is produced in a body**—as in bending a beam, or twisting a shaft, or compressing a block of any kind. Imagine a small load, say a pound weight, applied to a spiral spring; the spring elongates a certain amount and stops, showing that its ability to resist the elongating force gradually increases as it is stretched. If the load be doubled by adding a

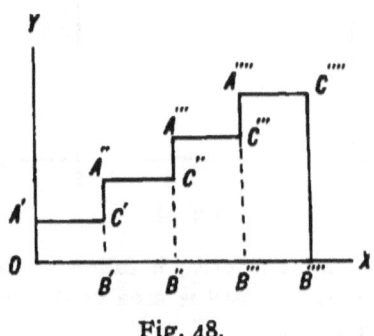

Fig. 48.

second weight the spring will stretch still further,—until the internal resistance again becomes equal to this external load. Now, if we measure the length of the wire under this load we will find that the elongation for the two pounds has been just twice as great as for one. If we add a third pound the elongation becomes three times as great, the internal resistance, of course, increasing in the same ratio in order to balance the weight. In other words, the internal resistance of the spring is directly proportional to the elongation. The same idea applies to all bodies, whatever the nature of the stress. This is called **Hooke's Law** which asserts that within certain limits the internal resistance of a distorted body is directly proportional to the amount of distortion or deformation. When the internal resistance becomes equal to the external force the distortion ceases; and hence for bodies at equilibrium under stress—the usual case—the law might be expressed in the converse form, viz., that the distortion is directly proportional to the distorting force.

In graphical form the work done in elongating the spring by adding successive weights of one pound each would be represented as in Fig. 48. Let OA' represent the force of one pound and OB' the elongation produced by this force; the work done thereby will be the area $OB'C'A'$. Then a second weight was added, making two pounds in all, as $B'A''$, and producing an additional elongation $B'B''$, or a total of OB'' equal to $2\,OB'$. As we go on stretching the spring farther and farther, the area representing the work increases step by step in the manner pictured in the diagram. The work done in

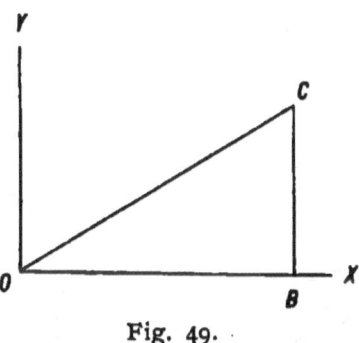

Fig. 49.

stretching the spring through the distance represented by $B'''B''''$ is four times as great as in stretching it through the same distance OB' at the first stage of the elongation.

If the tension had been applied to the spring more gradually, increasing continuously instead of a pound at a time, the diagram would have been as in Fig. 49. If the tension increases uniformly from zero to BC, while the spring is elongated an amount represented by OB, the work done will be represented by the area of the triangle OBC,—which is only half as great as if the maximum load BC acted throughout the entire distance, as in Fig. 47. Or, the average load is only a half of BC.

What is true in this respect for the elongation of a spring is true for a body distorted in any manner.

In the **Indicator Diagram** shown in Fig. 34, p. 47, each half of the diagram shows the changes of steam pressure in one end of the cylinder for a complete or double stroke of the piston. The area enclosed within $ABCDEA$, Fig. 50, represents the work done in one end of the piston, and the corresponding area $A'B'C'D'E'A'$ is the work done in the other end. The total work done by the steam for a complete stroke is the sum of the two. In the diagram the two areas overlap in part, but this is only a matter of convenience in the mechanical process of taking the card from the engine; in computing the work done this area is counted twice.

When the piston is at the right end of a stroke in the diagram, a small amount of steam is confined in that end of the cylinder under pressure $D'A$ to form a cushion, as explained on p. 78, while the other end at that instant is in communication with the air and hence

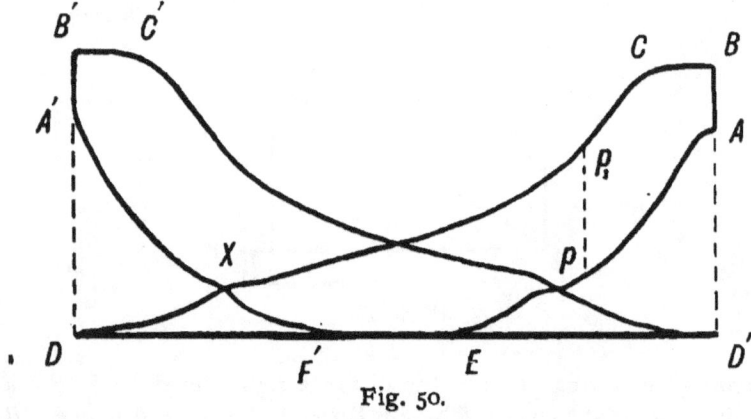

Fig. 50.

has no pressure above the atmospheric. In this position, steam is admitted to the right end and the pressure in that end instantly jumps from $D'A$ to the boiler pressure $D'B$. The piston moves to the left. The right port apparently remains open to the full boiler pressure for about one-tenth of the stroke, when it closes (at point C); for the balance of the stroke the steam works by expansion, gradually falling off from the full pressure at C to the atmospheric pressure at D. At D the piston completes the half stroke and starts back. The right end of the cylinder is now open to the atmosphere and remains so until point E is reached, when it closes for the purpose of retaining sufficient steam to form the cushion by being compressed from E to A. If it were not for this compression the work done by this end of the piston for a complete stroke would be represented by the somewhat triangular area $DD'BCD$. Out of this area we must subtract the area $ED'A$, lost by compression during the return stroke.

While this subtraction is mathematically correct it does not give a correct idea of the actual transaction. The work of compression represented by the area $ED'A$ is really done by the steam in the other end of the cylinder, and properly should be subtracted from the area $D'DB'C'D'$. The two halves of the diagram should be read together. The right end actually does the full amount of work represented by the area $DD'BC$, a part of this being used up in the work of compressing the hold-over steam in the other end of the cylinder (area $E'DA'$) and the balance being given to the engine for useful work. We must not only read the conditions in the two ends of the cylinder at the same time, but also in their relations, each to the other. As the pressure in the right end of the cylinder falls from B to D, the pressure in the other end changes from D' to B', the compression commencing at E'. From the intersection, X, of these two lines the pressure on the driving end of the piston is actually less than on the other end, and if it were not for the kinetic energy of the fly-wheel the piston would not complete its stroke but would bound back before reaching D.

Hence the area $DD'BCD$ *minus* area $E'DA'$ is the *effective* work done during a half revolution of the engine, and $D'DB'C'D'$ *minus* $ED'A$ is the effective work for the other half. Putting this in algebraic form we have

$$(DD'BCD - E'DA') + (D'DB'C'D' - ED'A)$$

as the total work done during a complete stroke or revolution. This can be transformed into

$$(DD'BCD - ED'A) + (D'DB'C'D' - E'DA'), \text{ or}$$
$$ABCDEA + A'B'C'D'E'A';$$

thus proving what was before stated, that this last result represents the actual mathematical value of the work done, even if it does not picture the transaction correctly.

The Mean Effective Pressure, referred to on p. 76, would be the average of an infinite number of lines drawn like pp_1 in the diagram. It is the combined areas $ABCDEA + A'B'C'D'E'A'$ divided by the length of stroke, DD'.

CHAPTER III.

CENTER OF GRAVITY.

Center of Figure. The Center of Figure of a straight line is at its middle point; as much of the figure lies on one side of the point as on the other. With equal facility we can locate the center of figure of a circle, an ellipse, a parallelogram, a sphere, a cylinder, or any other symmetrical figure. It coincides with the **center of symmetry**, as defined in geometry. The center of figure of a spherical shell is the same as if it were a solid sphere; a hollow box has the same center as a solid block of the same size and

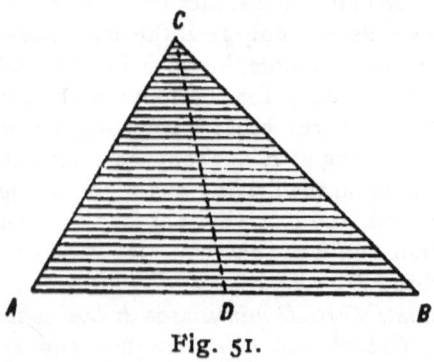

Fig. 51.

shape; likewise the center of a length of pipe is not in the material of the pipe but at the middle point of the axis of the enclosed cylindrical space.

Even if the figure is not strictly symmetrical, if it possesses some degree of regularity, there may be in it a point which may properly be called its center of figure. A triangle can be imagined to be made up of an infinite number of parallel straight lines, as in Fig. 51 and the center of each line will be in the median CD. Since all these lines taken together constitute the triangle, the center of the triangle must be somewhere in the median CD. If a second set of lines were drawn parallel to AC, or BC, their centers would all lie in a second median. The intersection of two of its medians is therefore the center of figure of the triangle.

Center of Mass. If a geometrically symmetrical body is also composed of uniformly distributed particles, the center of figure is the position of a central particle. And if, furthermore, the body is of a strictly homogenous substance or of uniform density, the central particle is also an average point about which the mass of the body is distributed,—or the center of mass, as it is commonly called.

If this body howsoever symmetrical in shape is not of uniform density the center of mass is not so readily located.

If a piece of lead is glued to the end of a cork, the central particle of the combined mass may or may not be at the center of figure, but the center of mass is evidently nearer the lead.

Center of Gravity. According to the law of gravitation, every particle of a body participates in the action of gravity, and, (if the body is small in comparison with the size of the earth, so that the lines from its different particles to the center of the earth may be regarded as parallel), the weight of the body is the sum-total of these minute attractions. For many purposes in mechanics, this downward tendency or weight of the body may be represented as a single force, acting at an average or central point called the Center of Gravity.

The importance of this conception is that the action of gravity, or any other force acting equally and in parallel directions on all the particles of the body, is the same in effect as if the entire mass of the body were condensed in its center of gravity, the remaining particles of the body being imagined as weightless.

Since the laws of mechanics deal with masses only through their relation to forces acting upon them, there is no discrimination that we shall need to exercise in using the terms "center of mass," "center of gravity" and "center of weight;" we may use one for another without danger of error.

To Locate the Center of Gravity of a Body. It has been shown that the center of figure of any symmetrical body can be found by simple geometric construction, and also that if its density is uniform its center of mass and center of figure will coincide. If it is a hollow body—a tube, box, ring or spherical shell—this central point is still called the center of mass, even though there is no central particle of the substance.

Having located the center of mass of each of several symmetrical bodies by geometric methods, we can use the results to prove the following

Proposition:—*If a body is supported freely and loosely on a pivot it will adjust itself so that the center of gravity will be vertically below (or above) the point of support.*

(a) Prepare two pieces of cardboard, one triangular and the other a parallelogram. Locate the center of each by geometry. In each make two pin holes near the edges and not too near each other. Prepare a plumb-line by tieing a screw, or other small object, to a fine thread. On a fine needle driven into the wall suspend the triangle from one of the pin holes, and hang the plumb-line from the same support. When both have come to rest, mark the lower edge of the triangle where it crosses the plumb-line. Remove from the pivot and draw a line from the point of support to the point marked on the edge. Now suspend the triangle from the second pin hole, and determine a second line. Do these lines intersect at the center of figure as determined by construction? If not, determine the amount of error by measuring the distance between the two locations. When the triangle was supported on the needle what was the position of the center of gravity relatively to the point of support?

In the same manner hang the parallelogram on the needle from each of the pin holes in succession, and note results.

(*b*) Remove the cover and bottom of a pasteboard box. Stick a pin in one of the edges so that the pin will run full length into the cardboard parallel to the side of the box. Tie the plumb-line at about its middle point around the pin just at its head. Fasten the top of the plumb-line to some convenient support and allow the bob to drop through the hollow of the box. When both have come to rest mark the lower edge of the box where it crosses the plumb-line. After removing the plumb-line measure the distance to the nearest corner from the pin hole and also from the marked point where the plumb-line crossed. Are these points located symmetrically?

By means of a needle and thread connect the middle points of the opposite sides of the box, thus locating the center of the box. Again suspend as before and see if the plumb-line passes through the center of figure.

These experiments will suffice to show that the center of gravity of a body ordinarily assumes a position *vertically* below the point of support. This is a simple fact that hardly needed demonstration; it is in perfect accord with an endless number of phenomena that we observe from day to day.

To support a body from a single point with its center of gravity vertically *above* the point of support—such as a cone balanced on its apex, or a parallelopiped on an edge—is theoretically possible, but in practice the most skillful equilibrist would require some small area—more than a mathematical line or point—for a base of support.

Equilibrium. (*a*) *When a body is supported from a pivot*, with its center of gravity vertically above or below the point of support, it is said to be in a position of equilibrium. The weight of the body, acting like a single downward force exerted at the center of gravity, is balanced or equilibrated by the resistance of the pivot. If the center of gravity were not in vertical line with the point of support, as in Fig. 52, the pivot could not furnish a resistance equal and opposite to the force exerted at *C* by the weight of the body. To assume a position of equilibrium the body must rotate around the pivot until the vertical *CD* passes through *P*.

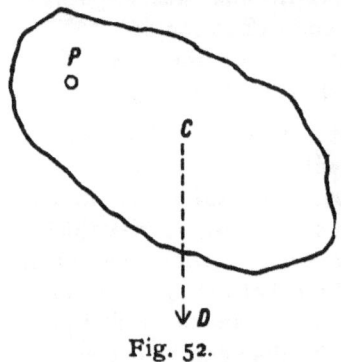
Fig. 52.

(*b*) Problems of equilibrium are more frequently met with in *bodies resting on a base of support;* pivoted bodies came first in order of consideration, not because of their

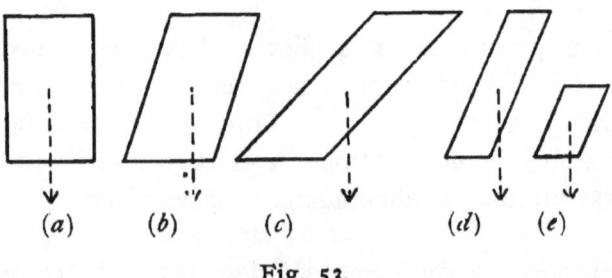
Fig. 53.

greater importance, but because it is simpler to deal with a single point of support than with an area of support. If a rectangular block rests on one of its faces on a horizontal plane (Fig. 53-*a*), it is in equilibrium, because the vertical

CENTER OF GRAVITY. 99

drawn through the center of gravity passes through the base of support, and the force of gravity is balanced by the resistance of the object supporting the base. If the block were not rectangular, as in Figs. 53-*b*, *c*, *d* and *e*, it would be in equilibrium or not, depending upon the area of base, the vertical height and the acuteness of the angles. Figs. *b* and *c* have the same area of base and the same vertical height, but the angles of *c* are such that the vertical from the center of gravity falls outside the base of support and the body is not in equilibrium. Figs. *b* and *d* have the same angles and the same vertical height, but Fig. *d* is not in equilibrium because of its smaller base. Figs. *c* and *d* would both be toppled over by their own weight.

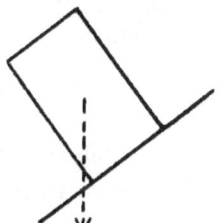

Fig. 54.

A body at equilibrium on a horizontal plane might not be at equilibrium if the plane were inclined, as shown in Fig. 54. A sphere is always at equilibrium on a horizontal plane, but is never so on an inclined plane.

If a body rests on several isolated points, as a surveyor's tripod, or a table, the base of support is the convex polygon that would be determined by winding a thread around the external points of support. There may be many other supporting points within the perimeter thus determined, but as they do not extend the area of support they do not add to the stability of the body as regards overturning.

Stability of Equilibrium—Stable; Unstable; Neutral.

(*a*) *Pivoted Bodies.* It has been stated that when a body is suspended freely from a pivot its center of gravity assumes a position vertically below or above the point of support; otherwise the body is not in equilibrium. One exception however, should be noted. If the point of support coincides with the center of gravity the body is at equilibrium in any position, and will remain at rest wherever placed by turning it around the pivot.

When the center of gravity of a pivoted body is vertically above the point of support its position is very insecure; any slight displacement of the body will cause it to roll over with its center of gravity downward. Though the body was in equilibrium it lacked stability, and is said to be in a position of **unstable** equilibrium.

But the position it naturally assumes, with the center of gravity vertically below the point of support, is the one of maximum stability. If it is displaced from this position it returns to it by force of its own weight. It is in **stable** equilibrium.

When the point of support coincides with the center of gravity the equilibrium of the body is **neutral.**

From these considerations, therefore, we derive two tests by which to judge the **kind of equilibrium** of a body.

(*i*) Its equilibrium is stable if the body tends to return to the same position after a slight displacement; but if a slight jar or disturbance causes the body to move still farther from its first position its equilibrium was unstable; if it remains wherever it is placed, its equilibrium is undisturbed by the change and is said to be neutral.

(*ii*) When the center of gravity is in its lowest possible position vertically below the point of support, the body is in stable equilibrium; when the center of gravity is at its highest possible position vertically above the point of support the body is in unstable equilibrium; when the center of gravity coincides with the point of support the equilibrium is neutral.

A third test of the stability of a body can be deduced from the consideration that any motion of the body around the pivot, which results in elevating the center of gravity requires the same amount of work—the same expenditure of energy—as if the whole mass were raised bodily through the same distance. If a bar of iron, shown in Fig. 55, weighs 8 pounds and its center of gravity is raised through a vertical height of 2 feet by rotating through the arc CC', then the work done is 16 foot pounds,—although a part of the rod has not been raised at all.

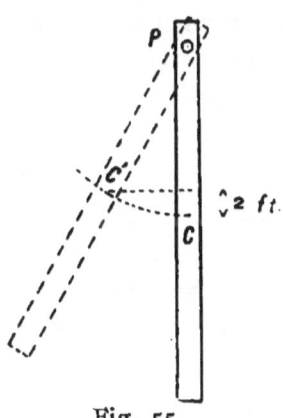

Fig. 55.

(*iii*) Now consider the three positions of the center of gravity relatively to the point of support— vertically below; vertically above; and coinciding with it—and the performance of work involved in any slight displacement from each of these positions of equilibrium. In the first case the center of gravity is raised and hence work is done *upon the body* to displace it from a position of stable equilibrium. In the second case the center of gravity is lowered, and hence when the body is displaced from a position of unstable equilibrium work is done *by the body*, its potential energy being converted into energy of motion. In the third case the displacement of the body is not accompanied by any change in the position of the center of gravity, and hence a body in neutral equilibrium has the same potential energy in all positions.

(*b*) **Bodies resting on a base.** A cone placed in various positions on a horizontal plane will afford typical illustrations of the three kinds of equilibrium. Resting on its base it is stable; balanced on its apex it is unstable; and lying on its side—on an element—it is in neutral equilibrium.

In general, if a body rests (at equilibrium, of course) on a base of any appreciable area it is stable, because any displacement tending to overturn it, or give it a new base of support, will result in elevating the center of gravity, and hence require an expenditure of energy. If it rests on a point or line, as a cube on a corner or an edge, with its center of gravity vertically above, in such manner that any displacement of overturning would lower the center of gravity, it is unstable. But if it rests on a point or line in such manner that the support may be shifted to other points or lines—a sphere, or a cylinder on its side—without raising or lowering the center of gravity, it is in neutral equilibrium.

If a body rests on an area of support with the center of gravity vertically above any point in the perimeter of the base, as in Fig. 56, we have a limiting case. Any slight displacement to the right will prove the present position of the body to be one of unstable equilibrium, while from a displacement to the left it will recover its present place as if it were stable.

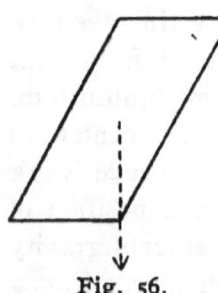
Fig. 56.

Examples:

 1. *If a rectangular block rests on one face what kind of equilibrium does it possess? Why? How must it be placed to be in*

CENTER OF GRAVITY. 103

unstable equilibrium? Could it be placed in a position of neutral equilibrium? Why?

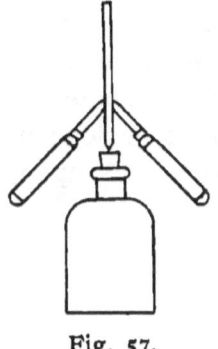

Fig. 57.

2. *How must a cylinder be placed to be in stable equilibrium? Could it be placed in a position of unstable equilibrium?*

3. *Can a sphere be placed in a position of stable equilibrium on a horizontal plane?*

4. *A pencil cannot be readily balanced on its sharpened point, but if a knife-blade be stuck into each side, it can then be balanced on its point from an elevated support, as in Fig. 57. Explain.*

5. *Fig. 53-e is stable, while Fig. 53-d, having the same base, is not stable. Explain.*

Degree of Stability. We have observed that a pivoted body has a choice of but three positions of equilibrium, and in each of these positions we regarded it as possessing a certain *kind* of equilibrium. These differences of kind we expressed by means of adjectives which signified merely the stability or non-stability of the body. In algebraic notation we could have said that in one position the body is positively stable to a certain degree, and in another position it is *un*stable or negatively stable. The numerical value of this degree of stability may be great or small, depending upon the weight of the body, and the distance between its center of gravity and the point of support. In a position of neutral equilibrium this distance vanishes and the stability is zero. It follows, therefore, that the equilibrium of a pivoted body may be represented in all its aspects by a single algebraic quantity, of which the numerical value will represent the degree of stability, and the sign of quantity will show the *kind* of equilibrium.

When a body rests on a base of support its degree of stability is measured by the work necessary to overturn it. For example, let us investigate the stability of a rectangular block resting on a horizontal plane. Suppose it measures 2 feet x 3 feet x 4 feet and weighs 150 pounds. There are six cases to be considered, since the work to be done in overturning depends upon which face the block rests upon, and the edge over which it is turned.

These cases are as follows :
1. Resting on 2 x 3 base.
    (a) Turned over 2-foot edge.
    (b) Turned over 3-foot edge.
2. Resting on 3 x 4 face.
    (a) Turned over 3-foot edge.
    (b) Turned over 4-foot edge.
3. Resting on 2 x 4 face.
    (a) Turned over 2-foot edge.
    (b) Turned over 4-foot edge.

Solution:
1. Resting upon a 2 x 3 face.
    (a) Turned on 2-foot edge.

If the body starts from the initial position $A\,B\,C\,D$ (Fig. 58) and is overturned on the two-foot edge, its center of gravity will describe the arc $O\,O'O''$. Since the edge AB is 3 feet and the height BC is 4 feet, the distance OA will be $\dfrac{\sqrt{3^2+4^2}}{2}$ or 2.5. When the body reaches the position $A\,B'C'D'$ it occupies a position of unstable equilibrium and no further work will be required for overturning; if it is displaced to the slightest degree beyond this position it will then fall by its own weight to the position $A\,B''C''D''$. Therefore, the work necessary to overturn the body from its first position of stable equilibrium is the amount required to raise the center of gravity from O to O',—which, as we have already learned, is equivalent to raising the entire mass through the same distance. In describing the arc $O O'$ the

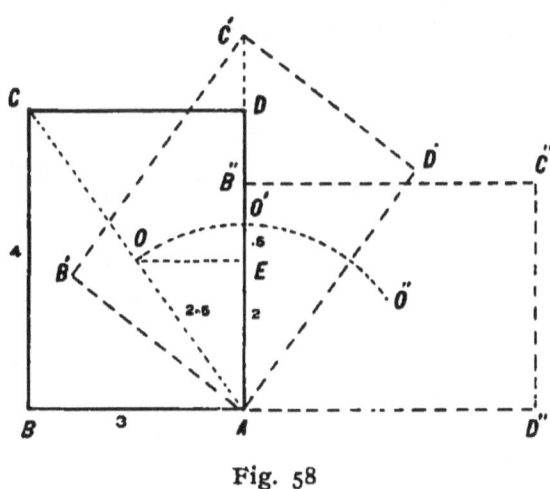

Fig. 58

center of gravity is raised through the vertical distance $E O'$, equal to 0.5 ft. The weight of the body being 150 pounds, the work done is 150×0.5 or 75 foot-pounds.

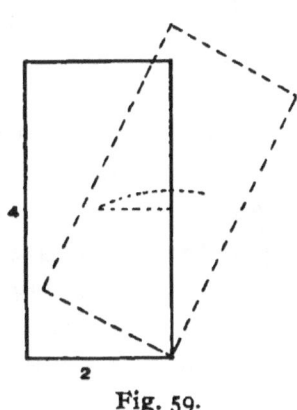

Fig. 59.

This gives us a measure of the stability of the body as opposing any effort to turn it from the given position, over the edge specified. It would be hardly sufficient, however, to say merely that the stability of the body is 75 foot-pounds; we must specify all incidental conditions upon which this depends,—the face upon which it rests, and the edge over which it is turned.

(*b*) Resting on the same base and turned on the 3-foot edge the vertical distance through which the center of gravity will be raised (Fig. 59) is 0.236 ft., and the work done is 35.4 foot-pounds.

The remaining cases are to be solved by the student and the results entered in the following table:

	Height of Center of Gravity in Feet	Width of Base in Feet	Stability
I $\dfrac{a}{b}$	2	3	75
	2	2	35.4
II $\dfrac{a}{b}$			
III $\dfrac{a}{b}$			

By comparing these results it will be seen that the body possesses the greatest stability when the center of gravity is as low as possible and the base as wide as possible.

This will be shown still more clearly if the results are re-arranged in proper places in the table on the following page, in which the width of base *in*creases in spaces from left to right, and height of center of gravity *de*creases from the top downward.

CENTER OF GRAVITY.

For any given base the stability is greater when the center of gravity is lower; for any given height of the center of gravity the stability is greater when the base is broader.

	WIDTH OF BASE		
HEIGHT OF CENTER OF GRAVITY	2 FEET	3 FEET	4 FEET
2 FEET	35.4	75	
1.5 FEET			
1 FOOT			

Examples:

1. *What work is done in overturning a cube of metal measuring 1 foot each way, and weighing 500 pounds?*

2. (a) *A mass of metal measuring 1 foot x 1 foot x 2 feet and weighing 1000 pounds rests on one end. What work is done in overturning it?*

(b) *What work will be done in overturning it from a position of rest on one side, turning it on the 2-foot edge?*

(c) *What if turned from the same position on the 1-foot edge?*

3. *A stove 3 feet high, 2 feet wide and 3 feet long, weighing 300 pounds, rests on legs 6 inches high. Another weighing 500 pounds, rests on legs 12 inches high and measures 3 feet high, 3 feet long and 3 feet wide. Which stove is the most stable, and in what ratio?*

NOTE: Disregard weight of legs and consider center of gravity as being at center of figure of the "body" of stove.

CHAPTER IV.

PRINCIPLES OF MACHINES. THE LEVER.

Tools and Machines. By means of a few hand tools a blacksmith overcomes the resistance of a piece of iron and forges it into almost any desired shape, in a manner that would not be possible by unaided human effort. Similar results are accomplished in any material, as wood or stone, by means of suitable tools, or with greater facility by woodworking and stoneworking machinery operated by power. Most tools and machines of this type, used for purposes of construction and including many of the highly specialized devices used in various factories, serve the purpose of cutting and shaping materials.

Under a different head we could classify machines that are used for lifting loads or moving masses of any kind. This class would include chain hoists, cranes, pumps, air blowers, elevators, etc., and in the same list we might even include a dynamo, in which the resistance overcome is something more than the mere mass of the moving parts and the results obtained are not so readily discerned.

A third class would be machines that are designed to accomplish delicacy and accuracy of motion, without overcoming any great resistance, and would include weaving machines, sewing machines, typewriters and type-setting machines.

This is by no means a complete list or classification of the many different kinds of tools and machines in use, but it is sufficient for the purpose of opening the way to a definition that will indicate the fundamental character or essential nature of a machine.

A machine is an instrument by which a given force is made to accomplish a result indirectly that would not be possible by direct application of the force without the intervention of such a medium. *It is a device that serves to modify a force or motion,* in magnitude or direction, or in both respects. Any tool, implement, device, contrivance, instrument, appliance, or apparatus, by which this is accomplished, directly or incidentally, involves a mechanical principle and is a machine, according to our definition. Animal motions, even, are due to the various mechanisms of which the anatomical structure is composed.

Efficiency of Machines. As already stated, many machines are intended to lift loads and in other ways overcome resistance. But even those which are designed to accomplish merely a delicate movement of any part in a certain direction cannot be operated except by the application of energy supplied from some source external to the machine. If the machine were without weight and its parts could move without frictional resistance, it would be capable of doing an amount of useful work exactly equal to the energy applied. But that is practically impossible; owing to the frictional resistance, air currents, etc., due to the moving parts of the machine, some of the applied energy is dissipated, or frittered away, in the form of useless motion, heat, sound, and at times even light and electricity. If a machine performs only 385 foot-pounds per second of useful work at the expense of one horse-power of energy (550 foot-pounds per second) applied to it, its efficiency is only 70 per cent. The efficiency of a machine depends upon its structural features and the condition of the bearing parts.

A complex machine may be more efficient in some parts than in others. A machine-shop planer, for instance, might be conveniently segregated into three parts, (1) the horizontal bed; (2) the driving pulleys and accessory parts conveying motion to the bed; and (3) the countershafting. The countershafting may convey to the machine proper 96% of the power taken from the main shaft; the driving parts of the planer may furnish to the bed only 82% of the power received from the countershaft; and of this the losses due to motion of the bed may leave only 88%. Hence the complete machine would furnish in useful work only 88% of 82% of 96% of the power taken from the main shaft; or its efficiency would be less than 70%.

The Simple Machines. Any mechanical contrivance, howsoever complicated, can be analyzed into certain elementary parts, commonly designated as the **Mechanical Powers**, or **Simple Machines.** These are:

1. The Lever, (including the bent lever and bell crank).
2. The Wheel and Axle, (sometimes classed as a lever).
3. Pulleys.
4. Inclined Plane.
5. The Wedge, (which may be included under the inclined plane).
6. The Screw, (which may also be included under the inclined plane).
7. The Toggle Joint, (sometimes called by analogy the "knee-joint," or "elbow-joint.")

All the elementary or component parts of any machine could be classified under these heads. The most complicated machine is nothing more than an assemblage of parts involving combinations and modifications of these elementary mechanisms.

The Lever. A lever is a rigid rod free to move about a single fixed point called the fulcrum. It is the simplest form of a machine, and from it we can deduce the **fundamental principles** and considerations **of all the mechanical powers.**

(*I*) **Principle of Virtual Work, or Virtual Velocities.** If a weight W (Fig. 60), suspended from one end of a rigid rod, is balanced by a force P acting vertically downward on the other end, or on any point beyond the fulcrum, the relative values of P and W necessary for equilibrium depend upon the distances a and b, or AF and BF.

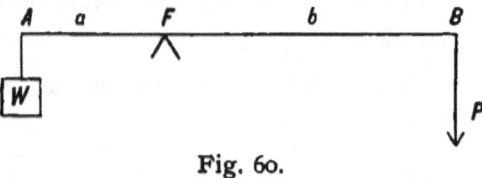

Fig. 60.

The two forces are inversely proportional to the distances from their points of application to the fulcrum.

$$\frac{P}{W} = \frac{a}{b}. \tag{9}$$

For example, if the short arm of the lever is 3 feet long and the other arm 5 feet, a weight of 10 pounds suspended from the end of the short arm would be balanced by a force of $10 \times \frac{3}{5}$, or 6 pounds, applied vertically downward at the other end.

This can be proved by the principle of Virtual Work. Assuming that the rod is without weight and that P and W produce equilibrium, suppose that the rod be given any displacement through an angle γ (Fig. 61), in either direction around the fulcrum. If the weight is raised to a position W', through a vertical height h, the other end falls through a vertical distance k. Now $h = a \sin \gamma$, and $k = b \sin \gamma$; whence $\frac{h}{k} = \frac{a}{b}$. If $h = 1$ foot, k is therefore equal to $\frac{5}{3}$ feet. Any displacement of the rod which raises

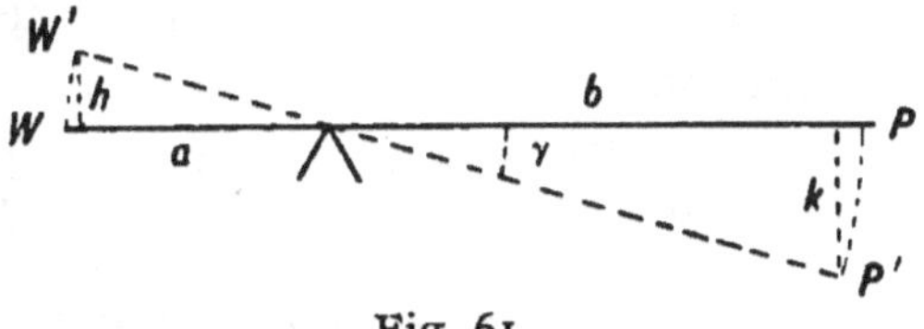

Fig. 61.

the weight one foot requires an expenditure of 10 foot-pounds of energy. But, as we have shown, this displacement of the weight would be accompanied by a falling of the other end, or a movement of the applied force, through a distance $\frac{5}{3}$ feet, and since the applied force is 6 pounds, the energy furnished by it is $6 \times \frac{5}{3}$, or 10 foot pounds. The work done by the applied force is therefore just equal to the work done upon the weight, and on the whole there is no gain or loss of energy. This is strictly in accord with the Law of the Conservation of Energy, and is applicable as a test of the equilibrium of any of the mechanical powers. That is, to determine whether the applied force is just sufficient to balance the weight or resistance, we imagine a slight displacement of the machine in such manner as it is free to move, and then calculate the product of the weight times the vertical distance through which it moves. This should be equal to the product of the applied force times its vertical distance; or, referring to the figure, we should have

$$Wh = Pk. \qquad (10)$$

This is called the **Principle of Virtual Work**, or **Virtual Velocities**. The adjective "virtual" is used to signify that there is no real motion or displacement, and no work is actually performed either upon the weight or by the applied force; but the result attained by the supposition that motion does take place, and that work is done, is virtually the same in effect as if the machine really moved, in spite of the fact that it is in equilibrium.

If $Wh = Pk$, then $\dfrac{W}{P} = \dfrac{k}{h}$. Since $\dfrac{k}{h} = \dfrac{b}{a}$, we have $\dfrac{W}{P} = \dfrac{b}{a}$, thus proving our proposition, that two balancing forces applied to a lever are inversely proportional to the distances of their points of application from the fulcrum.

(*II*) **Principle of Moments.** In the case under consideration, where the two forces act at right angles to the lever on opposite sides of the fulcrum, the distances a and b are called the lever arms. The product of a force times its leverage, or lever arm, is called the **moment of the force**. A force acting at the end of an arm 2 feet long has twice the advantage of the same force if its lever arm is only one foot.

The Principle of Moments asserts that two forces acting on a lever are in equilibrium if their moments are equal

PRINCIPLES OF MACHINES. 115

and tend to turn the lever in opposite directions. The demonstration of this principle requires only a simple transformation of the expression $\frac{W}{P} = \frac{b}{a}$ to the form $Wa = Pb$. In the case under consideration (Fig. 60) the weight 10 pounds, with a leverage of 3 feet is exactly balanced by the opposing force of 6 pounds with a leverage of 5 feet, because both have moments of 30 units.

A moment is usually designated as so many pound-feet, or ton-feet, for obvious reasons. These should not be confounded with the unit of work, the foot-pound. In both cases we use a compound word made up of a unit of force and a unit of distance, but the order of combining the component words is different in the two cases, and furthermore the unit of distance represents a leverage in one case while in the other case it refers to the distance through which a weight is raised or other resistance overcome.

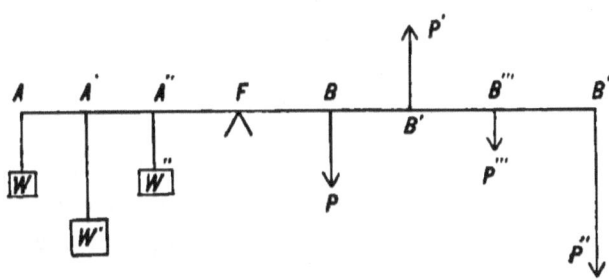

Fig. 62.

If several forces and weights act on the same lever simultaneously, the sum of the moments of the applied forces must be equal to the sum of the moments of the weights, to produce equilibrium. And if any of the forces act in a direction tending to aid the weights, such forces should be given a negative moment. In Fig. 62, the forces and weights are in equilibrium provided

$$W \times AF + W' \times A'F + W'' \times A''F = P \times BF - P' \times B'F + P'' \times B''F + P''' \times B'''F.$$

Moment Due to Weight of Lever. A rod of uniform dimensions supported midway between the two ends is in equilibrium, because its center of gravity is in vertical line with the point of support, (see p. 98). But if the fulcrum is not at the center, as illustrated in Fig. 63, the weight of the rod acts like a load applied at its center of gravity; as if the entire weight W of the rod were concentrated at that point. Hence, if such a rod were used for a lever, we

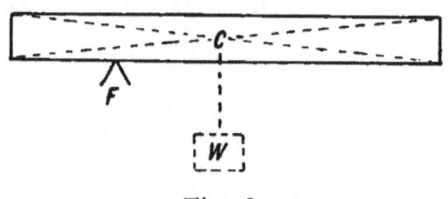

Fig. 63.

could allow for its weight by considering its effect equal to moment $W \times CF$. If the center of gravity is over the fulcrum, this moment is zero and the weight of the rod may be disregarded.

Sometimes it is easy to allow for the weight of the rod by proportioning the weights of the parts on opposite sides of the fulcrum, and multiplying the weight of each part by the distance of its center from the fulcrum, as illustrated in Fig. 64, in which W_1 and W_2 are the weights of the portions of the lever on opposite sides of the fulcrum. If the weight of each part is assumed to be massed at its center, then the heavy rod is equivalent to a weightless rod with the weights W_1 and W_2 suspended from C_1 and C_2 respectively.

This method seems to suggest itself to most persons more readily than the more direct method previously described, but it is a roundabout procedure and sometimes involves difficulties.

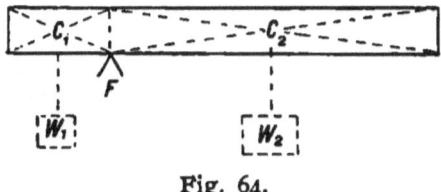

Fig. 64.

Examples:

1. *A heavy rod balanced on a fulcrum at its middle point has a weight of 40 pounds suspended from a point 2 feet from the fulcrum.*
    (a) *At what distance on the other side of the fulcrum must an 18-pound weight be placed in order to produce equilibrium?*
    (b) *What weight would have been sufficient to produce equilibrium if it had been placed 3.5 feet from the fulcrum?*

2. *A heavy rod 3 feet long is balanced on a fulcrum at its middle point. If a weight of 10 pounds is suspended from one end and a second weight of 5 pounds is placed at a point one foot from the fulcrum on the same side, what downward pull must be exerted on the other end of the rod to produce equilibrium? What if the 5-pound weight had been placed one foot from the fulcrum on the side towards the downward pull?*

3. *If a stick of timber 9 feet long and weighing 16 pounds is supported on a fulcrum 3 feet from one end, what weight must be suspended from this end to produce equilibrium? Where would a 12-pound weight be placed to produce the same result?*

4. *A uniform iron rod 11 feet long and weighing 40 pounds is supported on a fulcrum 3.5 feet from one end. If a 14-pound weight is hung from this end, where must a second weight of 28 pounds be placed to produce equilibrium?*

5. *For the Principle of Work we found that* Wh = Pk, *and for the Principle of Moments* Wa = Pb. *Explain the difference between these principles by contrasting the meanings of* h *and* k *with* a *and* b.

THEORETICAL MECHANICS.

Three Kinds of Levers. The applied force and the load acting on a lever are not always on opposite sides of the fulcrum. If a rod is fixed at one end, as shown in Figs. 65

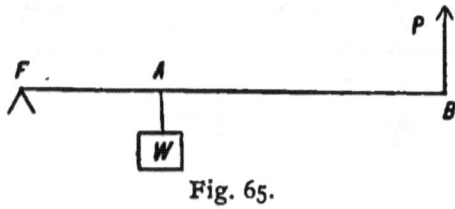

Fig. 65.

and 66, a weight acting at any point may be balanced by a force exerted at any other point. The relative values of P^* and W^* depend upon the distances AF and BF, exactly

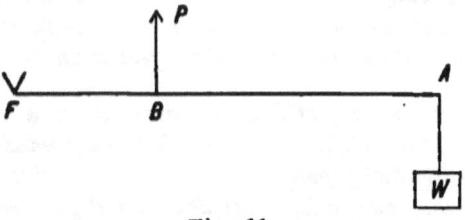

Fig. 66.

as determined for the lever of the first kind, from which we deduced the Principle of Work and the Principle of Moments. Whatever the relative positions of P, W and F, these principles apply with equal strictness. In Fig. 65, since $W \times AF = P \times BF$, and BF is greater than AF, the applied force is less than the weight. In Fig. 66 it is greater.

---

*The employment of the letters P and W, to represent respectively the Applied Force and the Weight, has become a matter of conventional usage. It is assumed, of course, that W does not always stand for a weight or gravitational action, but may be a load, or a tension, or resistance of any kind whatsoever. The applied force P is frequently called the "*Power;*" whence the symbol P. This use of the word "Power," referring to a force merely, is not consistent with our previous acceptance of its meaning in the sense of "Rate at which a machine can do work." For the sake of accuracy we shall adhere to the expression "Applied Force," symbolized by the capital P.

If P and W are on opposite sides of the fulcrum, the lever is said to be of the "first order." Fig. 65 illustrates a lever of the "second order," and Fig. 66 one of the "third order."

In levers of the first order the applied force may be greater or less than the load. In levers of the second order the applied force is always less than the load, while in levers of the third order the load is less.

Compound Levers. In some devices — certain gate latches and wagon brakes, for examples,—a train of levers is used to multiply the force or motion, as the case may be.

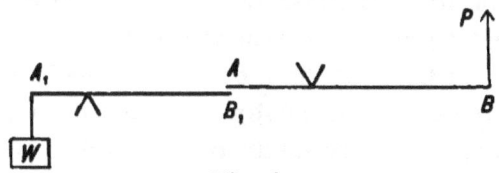

Fig. 67.

As shown in Fig. 67 the weight, or resistance, of one is propagated to the next as an applied force. It is not necessary that all in the series be levers of the same order.

Examples of compound levers are found in platform scales, typewriting machines, piano keys, the trigger of a gun, and in railroad switches.

Examples:

1. *To which class of levers would you assign each of the following devices:*
   Pair of pliers; sugar tongs; scissors; nut cracker; blacksmith's vise; blacksmith's tongs.

2. *Name five devices in which the lever is used. At least one of the five must refer to the lever of the third order.*

3. *How many of the three kinds of levers are illustrated in the ordinary use of a crowbar?*

Safety Valves. The ordinary safety valve, used as a boiler attachment to prevent explosions, is a lever of the third order. A valve V (Fig. 68) in direct communication with the interior of the boiler, is pivoted at B to a bar AF.

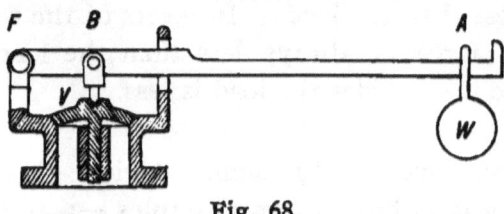

Fig. 68.

Whenever the upward pressure on V reaches a certain limit sufficient to overcome the resistance due to the weight of W and of other parts of the mechanism, the lever and valve are moved upward, turning on F as a fulcrum. By this means a part of the contents of the boiler is allowed to escape, and if by that means the boiler pressure is sufficiently reduced, the valve drops back into its seat.

Unless the area of the valve is sufficient to allow the steam to escape as fast as it is generated, perfect safety is not secured. As the generating capacity of the boiler depends mainly upon the area of the grate surface, it is customary to allow one square inch of valve area for every two square feet of grate surface.

Examples:

1. *How would you find the total upward pressure on the valve? Should you allow for the atmospheric pressure on the top of the valve?*

NOTE: The reading of a steam gauge indicates, not the actual boiler pressure, but the difference between this and the pressure of the atmosphere.

PRINCIPLES OF MACHINES. 121

2. *What pressure per square inch (by gauge) would be necessary to raise a safety valve constructed and adjusted as follows:*

> Weight of valve, 8 pounds.
> Diameter of valve, 3 inches.
> Valve pivoted 4 inches from fulcrum.
> Weight of lever 12 pounds (uniform rod).
> Total length of lever, 2 feet.
> Weight of 150 pounds suspended from extreme end of lever.
> All as illustrated by diagram, Fig. 69.

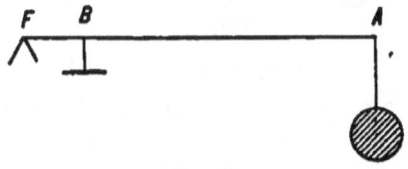

Fig. 69.

3. *At what point on this lever arm would it be necessary to place a 200-pound weight to withstand the same pressure?*

4. *At what point must the 150 pound weight be placed to balance a boiler pressure of 100 pounds per square inch (by gauge)?*

5. *What weight must be placed at the end of the rod to balance a boiler pressure of 90 pounds per square inch (by gauge)?*

6. *If a boiler is over a grate 5 feet square, what diameter of safety valve should it have?*

Pressure on the Fulcrum—Parallel Forces. (*a*) Levers of the First Order. If a ball weighing 10 pounds (Fig. 70) is fastened to one end of a weightless lever, 3 feet from the fulcrum, it can be balanced by a second ball weighing 6 pounds and fastened on the other end at a distance of 5 feet from the fulcrum. It is obvious, in this instance, that the fulcrum bears the entire load of 16 pounds, $W + P$. And since the entire system of combined masses is at equilibrium its center of gravity must be in vertical line with the point of support, as if the entire weight (16 pounds) were acting downward on that point.

Since the distances of the fulcrum from A and B are inversely as the weights acting at those points, it follows as a general **principle** that if two **parallel forces** in the same

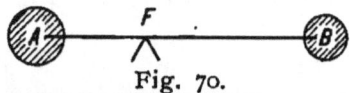
Fig. 70.

direction act on two different points of a body, their resultant is equivalent to a single force equal to their sum and acting at a third point whose distance from the two points of application is inversely as the forces.

If more than two parallel forces are acting simultaneously on the same body, the resultant of any two may first be found, and this may then be combined with a third force, etc.

Examples:

1. *A rectangular block of wood 1 cm. x 1 cm. x 2.5 cm. is glued to a cube of lead 1 cm. x 1 cm. x 1 cm., in the manner shown in*

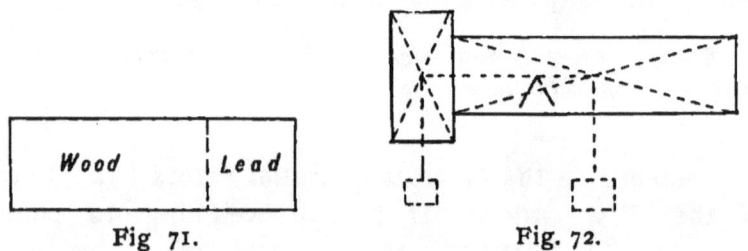

Fig 71.   Fig. 72.

Fig. 71. Locate the center of gravity of the combined masses. (Specific gravity of lead = 11.3; specific gravity of wood = 0.6).

2. *A ½-inch bolt has a head ⅞ inch square and 7/16 inch high. If the shank of the bolt is 2 inches long, locate the center of gravity of the entire bolt.*

PRINCIPLES OF MACHINES. 123

(*b*) Levers of the Second Order. In a lever of the second order the pressure on the fulcrum is $W-P$. For example, if a weight of 12 pounds is suspended at a distance of 1 foot from the fulcrum, it can be balanced by a

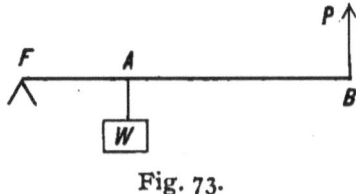

Fig. 73.

force of 4 pounds if the latter is applied at a distance of $BF = 3$ feet. In this case the fulcrum supports a load of 8 pounds.

To make this clear, imagine that the stick is supported by two persons, each having an end of the stick resting on one of his shoulders, and the 12-pound weight being suspended as stated, one foot from one end and two feet from the other. The person holding the end B will support 4 pounds, leaving the person at F to hold 8 pounds. Because, if the person at B should raise his shoulder the lever would turn about F as a fulcrum, and as the point B would describe an arc with radius FB, while A moves with radius $FA = \dfrac{FB}{3}$, then by the principle of work the end B would be lifted upward by a force $P = \dfrac{W}{3}$, or 4 pounds. If, on the contrary, the person at F should raise his end of the rod, moving it about B as a fulcrum, the arcs described by F and A would have radii of 3 feet and 2 feet, respectively, whence the upward force exerted at F would be $\dfrac{2}{3} \times W$, or 8 pounds.

If we choose to apply the principles of parallel forces, we can say that whatever the resistance which the fulcrum is required to exert it is equivalent to a force P' (Fig. 74) acting vertically upward. Then if the rod is in equilibrium the resultant of the two parallel forces P and P' must be equivalent to the single force of 12 pounds acting at point

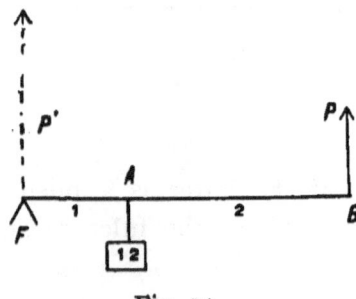

Fig. 74.

A,—a force equal to W but opposite in direction. And the moments of P and P' about A must be equal, or $P' \times 1 = 4 \times 2$, whence $P' = 8$.

(c) **Levers of the Third Order.**

Exercise:

Prove that in levers of the third order the pressure on the fulcrum is exerted in the direction of the applied force, and is equal to P—W.

Parallel Forces. The Couple. As a rule, but not always, two parallel forces acting on a body at different points can be counterbalanced by a third force. If the two parallel forces have the same direction, as P and P_1 in (Fig. 75), the resultant is a force R equal to their sum and acting at a point F such that $BF : AF :: P_1 : P$.

PRINCIPLES OF MACHINES.   125

This case is illustrated in levers of the first order, as was fully explained on pp. 121 and 122. If P and P_1 are equal, F is midway between A and B, as exemplified in a beam balance having equal arms.

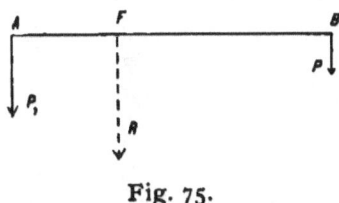

Fig. 75.

If the two forces are parallel and in opposite directions, as in Fig. 76, the resultant is a force equal to the difference between the two and acting at a point F, in the line AB continued, such that

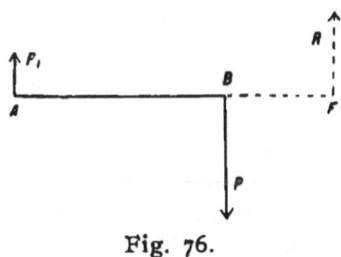

Fig. 76.

$AF : BF :: P : P_1$. If P_1 had been greater than P the point F would have been on the side next to A,—in BA continued instead of AB. This applies to levers of the second and third orders, pp. 123 and 124.

If P and P_1 are **equal** and **opposite** they have no real resultant. Two such forces acting on a body at different points constitute what is called a **Couple**. No single force howsoever applied could exactly counterbalance them, and hence they have no resultant. This will become evident if we observe the changes that take place in Fig. 76, as we substitute different values for P and P_1, finally making them become equal to each other. From the relation $AF : BF :: P : P_1$ it will be seen that, if P_1 in Fig. 76 should become smaller or P become larger, the point F will in either event approach nearer to B.

And conversely, if P becomes less, relatively to P_1 —that is, if P and P_1 become more nearly equal to each other—the point F will recede from B. A few simple computations will show that when P_1 finally becomes equal to P the point F recedes to an infinite distance. For that purpose let us substitute a few assumed values in the expression

$$\frac{AF}{BF} = \frac{P}{P_1}.$$

If $P = 2 P_1$, $AF = 2 BF$,   or $BF = AB$*
If $P = 3 P_1$, $AF = 3 BF$,   or $BF = 0.5 AB$
If $P = 5 P_1$, $AF = 5 BF$,   or $BF = 0.25 AB$
If $P = 10 P_1$, $AF = 10 BF$,   or $BF = 0.11 AB$
If $P = 100000 P_1$, $AF = 100000 BF$, or $BF = 0.00001 AB$

If $P = 1.5 P_1$, $AF = 1.5 BF$,   or $BF = 2 AB$
If $P = 1.1 P_1$, $AF = 1.1 BF$,   or $BF = 10 AB$
If $P = 1.01 P_1$, $AF = 1.01 BF$,   or $BF = 100 AB$
If $P = 1.0001 P_1$ $AF = 1.0001 BF$, or $BF = 100000 AB$

If the fractional difference between P and P_1 becomes infinitesimally small, then BF becomes infinitely greater than AB.

In some branches of applied mechanics the idea of the Couple is frequently met with, although we shall not need to use it in developing our subject from an elementary standpoint. It is illustrated in turning the handles of a copying-press, if equal forces are exerted on the two ends; likewise in cutting a thread on a pipe or bolt.

On account of its peculiar mathematical conditions the Couple possesses some very characteristic and striking properties. The sum of the moments of the two forces involved in a given couple is a constant quantity, being the same with reference to any point that may be selected, whether within or without the body acted upon by the couple, and if the body be pivoted so as to move around this point under the influence of the couple, the latter will cause no pressure whatsoever on the pivot or fulcrum.

---

* Combining $AF = 2 BF$ with $AF = AB + BF$ (Fig. 76), we get $2 BF = AB + BF$, or $BF = AB$. Solve the other cases in the same manner.

Bent Levers. If AFB (Fig. 77 or 78) is a rigid body free to turn about point F, the forces P and W will be in

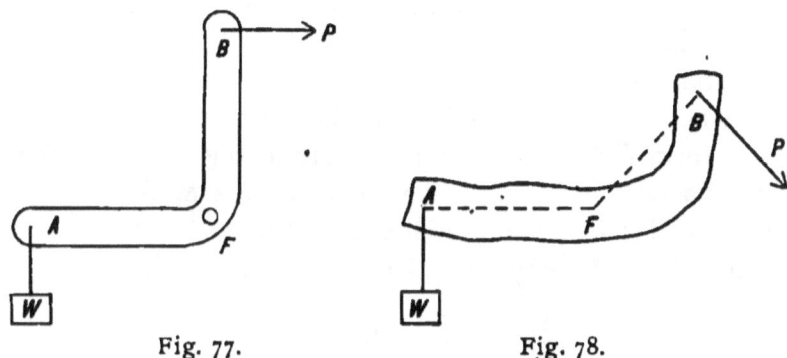

Fig. 77.   Fig. 78.

equilibrium if $W \times AF = P \times BF$. No matter what the shape of the lever, nor where the fulcrum is situated, if PB and WA are perpendicular to BF and AF respectively, the principle of moments applies exactly as if AF and BF were in the same straight line.

Moment of a Force Acting Obliquely on a Lever. Referring to Fig. 79, the moment of P with reference to the fulcrum

Fig. 79.

F is not $P \times BF$. By resolving P into two components, p_1 and p_0 (Fig. 80), perpendicular and parallel to BF, it

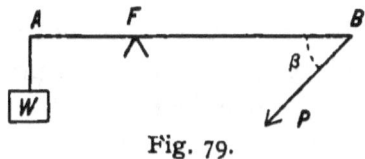

Fig. 80

becomes apparent that the component p_0 cannot have any effect in turning the lever; it simply pushes the lever in the direction BF against the bearings at F, without producing rotation. The component p_1, at right angles to FB, is alone effective in balancing W. Whence, if the lever is at equilibrium $W \times AF = p_1 \times BF$.

If the angle PBF is called β, then the component $p_1 = P \sin \beta$, whence $W \times AF = P \sin \beta \times BF$.

Instead of resolving the force P into two components, we might have turned to the idea of the bent lever by projecting the arm FB into a direction at right angles to P. The force P acting at an angle β with a lever arm BF has the same effective moment as if it were acting at right angles to the arm $B_1 F$, in Fig. 81. For, since $FB_1 = FB \sin \beta$, it

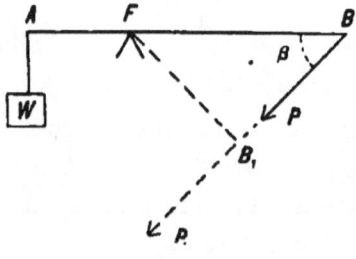

Fig. 81.

must follow that $W \times AF = P \times FB \sin \beta$. Therefore, when any force acts obliquely upon a lever, the effective moment of the force may be found in either of two ways: (1) By resolving the force into two components perpendicular and parallel to the lever arm; or (2) by projecting the lever arm into a direction perpendicular to the force. In either case the effective moment of the force is $P \times BF \times \sin \beta$.

It should be noted that the forces exerted to produce equilibrium in a bent lever, and likewise oblique forces on a straight lever, will give rise to a pressure on the fulcrum that is not merely the sum or difference of P and W. In Fig. 79 for example, a part of P (component p_0, Fig. 80) pushes the lever in the direction BF. The pressure on F, in this instance would be the resultant of $P_1 + W$ acting vertically and p_0 acting horizontally.

PRINCIPLES OF MACHINES.

From these considerations of oblique forces it follows that the **Principle of Moments**, deduced on pp. 114 and 115 for parallel forces, has a much more general application than was there assumed. If any number of forces in the same plane act upon a body to produce equilibrium, the sum of the moments of these forces about any point in that plane is equal to zero. The point with reference to which the moment is taken need not be within the body, and the forces need not be parallel, nor is it necessary in our diagrams to represent the form or outline of the body; the single line that we have used to connect the points of application of the forces is sufficient for all necessary computations.

In the lever shown in Fig. 82 the fulcrum is called upon to offer a resistance of 16 pounds, which is equivalent to a force

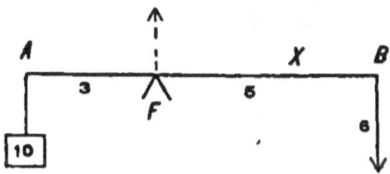

Fig. 82.

of 16 pounds acting upward, as if a string tied at F were pulled upward by such a force. Now, if we take any point as A the moments of the three forces, 6, 16 and 10 with reference to this point will be*

$$6 \times 8 - 16 \times 3 \pm 10 \times 0 = 0.$$

With reference to F the sum of the moments will be

$$6 \times 5 - 10 \times 3 \pm 16 \times 0 = 0.$$

With reference to a point X two feet from B the sum of the moments will be

$$6 \times 2 + 16 \times 3 - 10 \times 6 = 0.$$

---

*A force tending to produce motion in the direction of the hands of a clock around the point with reference to which the moment is being found is considered as having a positive moment; counter-clockwise a negative moment.

Examples:

1. *Referring to Fig. 82, select any other point in the rod and show that with reference to this point the sum of the moments of the forces involved is zero.*

2. *Prove the same for a point at any given distance beyond either end of the rod.*

3. *Take a case of a couple of any given numerical value, and find the sum of the moments of the two forces with reference to several different points,—say, one point in the line connecting the two forces; one in this line continued; and a third one at some convenient place entirely without this line. Is the sum of the moments the same in the three cases?*

If the two forces act on a body to produce equilibrium the sum of the moments of the forces is zero for any point and for all points. Can a couple produce equilibrium?

Mechanical Advantage. If a lever is used to enable a force of 6 pounds to balance a resistance of 10 pounds, there is an obvious advantage, but there is also the disadvantage that if motion takes place the load will not be moved as far as the applied force. Conversely, if a machine is contrived so as to multiply the motion, as would be the case if a force were applied to the short end of a lever of the first order to move a load at the longer end, and always in levers of the third order, then the motion gained would be at the expense of force exerted, for the applied force would then have to be larger than the load in proportion as its leverage is less. We can choose either advantage but we cannot accomplish both in the same machine; as already shown (pp. 114) $Wh = Pk$, so that the applied force and the load cannot be otherwise than inversely proportional to the distances through which they would move.

Some mechanical devices merely change the *direction* of the force or motion, without gain or loss of either. Take, for examples, the bell-crank and the single pulley, or even the lever with equal arms.

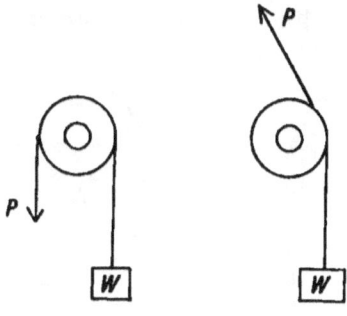

Fig. 83.

The bell crank shown in Fig. 77, changes the direction of the force ninety degrees, but if the two arms are equal, the force exerted is unchanged in magnitude.

The single fixed pulley (Fig. 83) permits no possible gain or loss of force, except by friction, but it can be used to accomplish any change of direction.

CHAPTER V.

MACHINES.

THE WHEEL-AND-AXLE. A pilot-wheel will serve as a convenient type of the numerous devices in which the idea the wheel-and-axle is employed. A comparatively small force applied to the handles on the periphery of the wheel is sufficient to overcome a greater resistance on the part of the tiller-ropes.

A load W applied at the axle or drum represented by the small circle in Fig. 84 has a moment of $W \times r$ with

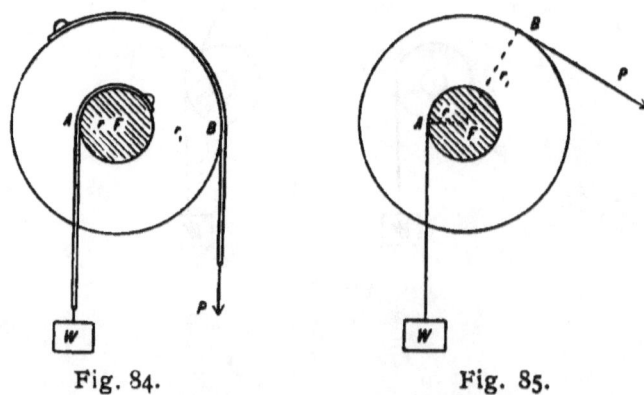

Fig. 84. Fig. 85.

reference to the center F. If the "wheel" has a radius r_1, the applied force will be just sufficient to produce equilibrium if $P \times r_1 = W \times r$.

As a statical machine—one in which it is desired to maintain a condition of rest and equilibrium—the wheel-and-axle is no more useful than a simple lever. Even if P were applied at a point on the periphery of the wheel such that A, F and B are not in the same straight line, as shown in Fig. 85, the result would be the same if we had used a bent lever. The wheel-and-axle and the lever are thus associated with each other, because, for purpose of computation, the principle of moments is readily applicable to both. But as useful machines they present this difference, that the wheel-and-axle permits of continuous motion, while the available motion of the ends of a lever is limited to the small distance resulting from an angular change of less than 90°

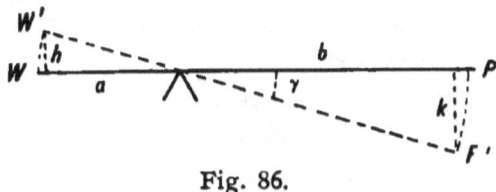

Fig. 86.

around the fulcrum. When the lever has turned through an angle γ (Fig. 86), if the directions of P and W have remained unchanged, their moments will become $Pb \cos \gamma$ and $Wa \cos \gamma$, and are zero when $\gamma = 90°$.

When the wheel-and-axle is used, P and W do not follow the points B and A to a new position (Fig. 87), as rotation occurs, but are shifted to new points of application, in such manner that the angles WAF and PBF (Figs. 84, 85 and 87), may be kept equal to 90°, — the position of maximum moment.

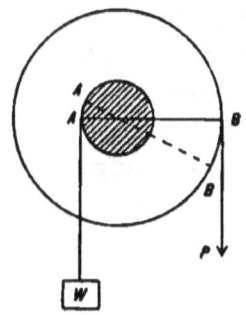

Fig. 87.

It is not always convenient, however, to preserve this favorable direction of P and W with reference to their lever arms. In many contrivances, like the windlass, the wheel is replaced by a crank, which acting like a single lever loses the advantages of a wheel. The rope of a windlass is always at right angles to its instantaneous lever arm in the axle, but the force applied to the handle attached to the crank is not always exerted to the greatest advantage, being sometimes at right angles to the crank and at other times acting at a less favorable angle. This is especially true of the crank of an engine, upon which the connecting rod acts at a varying angle. The spoke of a capstan is analogous to the crank of a windlass in construction, but a person operating a capstan, by walking in a circle, is always pushing at right angles to the lever arm, and hence nothing is lost by not having the entire wheel.

When a rope or belt acts on the circumference of a wheel an error is introduced in computing the moment unless we add to the radius of the wheel one half the thickness of the rope or belt, for, if the rope is wrapped around the circumference sufficiently to prevent slipping, it moves entirely with the wheel and the force is distributed equally over the cross-section of the rope or belt. Perhaps this can be made a little clearer by the principle of work. As long as the rope is coiled around the wheel the inner strands are compressed, and the outer portions elongated, beyond their normal condition. If a force is exerted on the rope causing the wheel to turn, the distance through which the force moves is the length of rope unwound, after the compressed and elongated portions have returned to their normal condition. Hence, in determining the work done by the force in moving the wheel through any angle, we take as the distance through which the force moves an arc described with a radius equal to the radius of the wheel *plus* half the thickness of the rope.

Exercise:

Give five examples of the wheel-and-axle, and if any involve the use of a crank, state whether the force acting on the crank maintains the position of maximum moment.

Gearing and Shafting. The various combinations of shafts and pulleys, used so commonly in the tramsmission of power, operate by a propagation of motion and force from one wheel-and-axle to another. The action of the entire system is traced from part to part in the same manner that a series of compound levers would be analyzed into its component simple levers.

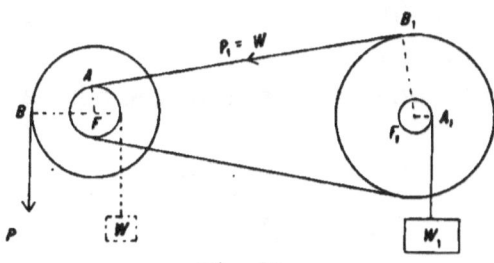

Fig. 88.

For example, if a force of $P = 90$ lbs. (Fig. 88) is applied at the circumference of a 24-inch pulley fastened on the same shaft with a 9-inch pulley, it will be equivalent to an opposing force of $W = \frac{24}{9} \times 90$, or 240 lbs., applied to the small wheel. Now, instead of moving the weight W, suppose it is desired to convey this action to a second wheel-and-axle and at the same time to modify it somewhat. If it suits our needs we may use a 30-inch pulley for the larger of these wheels and a 6-inch pulley for the smaller. By connecting the 9-inch pulley of the first wheel-and-axle with the 30-inch pulley of the second by means of a belt, these two wheels may be constrained to move with the *same lineal velocity*, and the force of 240 lbs. at the perimeter of the 9-inch wheel is exerted through the belt and thus applied to the perimeter of the 30-inch wheel. This force at B_1 will balance a load of $W_1 = \frac{30}{6} \times 240$, or 1200 lbs. at A_1.

Exercise: *Verify this result by the principle of work.*

When the speed is geared down to such an extent that very large forces are involved, it might not be practicable to use leather belting in the ordinary manner, because the belt would slip on the pulley before such forces could be exerted. If a rope is used it can be coiled around each pulley as many times as may be necessary to make it hold without slipping, (provided, of course, that the rope is stout enough to stand the tension). Or, if the shafts can be placed close to each other the power can be transmitted from one to the other by means of cog-wheels, as shown in

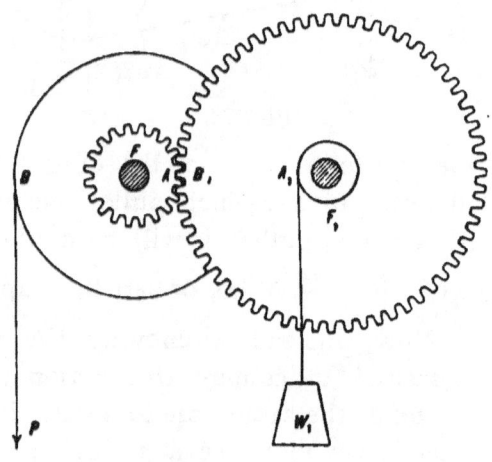

Fig. 89.

Fig. 89,—or by means of bevel gears if the shafts are not parallel. Since the number of teeth in the two cog-wheels is proportional to their circumferences, the computed relation between P and W_1 will be the same as if we had used the belt, but the possibility of slipping is obviated. Notice, however, that the direction of rotation of the second shaft is opposite to what it would have been if the transmission had been by belt.

From this figure the analogy between a system of gearing and a system of compound levers is readily observed. If BFA were a lever, instead of mere points in a wheel-and-axle, the downward force at B would cause an upward motion at A. From A the action would be transmitted to B_1 and the upward force on B_1 would have a lever arm $B_1 F_1$; etc.

This holds true whether the transmission is by belting or by spur gearing. The only probable source of error in either case would be in misunderstanding the real relation between the 30-inch and 9-inch wheels; although these pulleys may be of different diameters, they do not give rise to the same mechanical considerations as if they were on the same shaft. Two pulleys fixed on the same shaft constitute a wheel-and-axle, in which case they would have the same *angular* velocities, and hence would have different lineal velocities for points on the perimeters. But being mounted on different shafts and one taking motion from the other at their perimeters, as in the instance under consideration, they have the same *lineal* velocity, (but different rates of rotation). Therefore, in accordance with the principle of work, the action is propagated from the 9-inch wheel to the 30-inch wheel without gain or loss of force.

THE PULLEY. In machine shop practice the word "pulley" is employed to designate a wheel over which a belt runs in a system of shafting, as explained under the wheel-and-axle. Long before the transmission of power by belts and ropes, the wheel was used in a somewhat different manner as one of the simple mechanical powers, and it is in connection with this manner of usage that we apply the generic term The Pulley. Sometimes the mechanical advantage of the Pulley consists only in a change of direction,—a force applied in some convenient direction being employed to overcome an equal force acting in any other direction; under other circumstances, by a different arrangement of conditions, it may be made to modify the force or motion. Or, it may afford both of these advantages, especially when several wheels are compounded in the same machine.

Fixed Pulley. Very frequently a single pulley, arranged in the manner shown in Fig. 90, is used for purposes of hoisting. A grooved wheel is pivoted in a framework, which is suspended from a rigid support. In the matter of equilibrium between P and W, friction in this instance is an important consideration. But disregarding this, it is

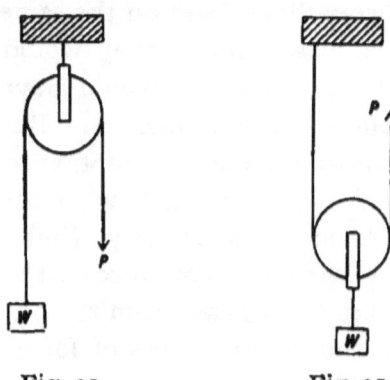

Fig. 90. Fig. 91.

obvious that according to the principle of virtual work $P = W$.

This is still true if the applied force is exerted in any direction, other than vertical; or even if the free end of the rope is carried by means of other *fixed* pulleys to any distance where it may be convenient to apply the action.

Exercise:

In the case of the fixed pulley referred to in Fig. 90, in which both P and W act vertically downward, what total load is the rigid support required to sustain?

MACHINES.

Movable Pulley. If a single pulley is arranged in such manner that one end of the rope is fixed to the support, while the pulley itself is free to move with the weight, the relation between P and W is greatly changed. Also, in this case, allowance has to be made for the weight of the pulley as well as for friction. But assuming a weightless pulley, P will be equal to $\dfrac{W}{2}$, because if motion takes place the free end of the rope will move twice as fast as the weight. To make this clear, imagine that the entire pulley and weight were grasped in the hand and raised to a height of one foot. The rope would then be left dangling through a vertical length of one foot on each side, and to take up the slack, the free end, to which P is applied, would have to be raised two feet.

Looked at as a simple question of statics, it will be observed that the total weight is supported equally by the two ends of the rope.

If the two parts of the rope are not parallel, the relation $P = \dfrac{W}{2}$, just deduced for the single movable pulley, is no longer true. If force is applied in a direction that is not parallel with the direction of W, the pulley and weight will not remain vertically below the point of support, but will roll to one side until they reach a position where the two parts of the rope make equal angles with a vertical line through the center of gravity of the pulley and weight (Fig. 92). The two parts of the rope will still be under equal tensions, and the load P_b on the point of support, in the direction of the rope on that side, will still be equal to

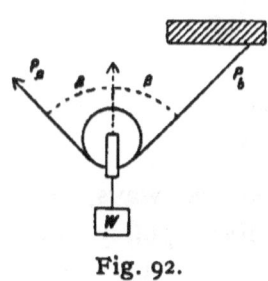

Fig. 92.

the applied force P, but P will not be equal to $\dfrac{W}{2}$. We must treat P_a and P_b as two component forces, and to produce equilibrium, their resultant must be equal and opposite to the known force W. Hence knowing the values of W and β, and knowing that the two components P_a and P_b are equal, we can compute the value of the applied force from the relations of the parallelogram (Fig. 93). If R represents this resultant, equal to W, then
$$R = 2\,P_a \cos\beta;$$
whence
$$P_a = \frac{W}{2\cos\beta}.$$

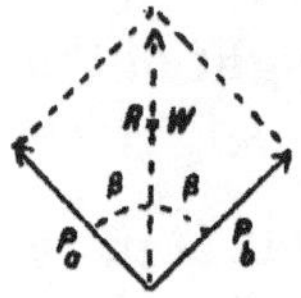

Fig. 93.

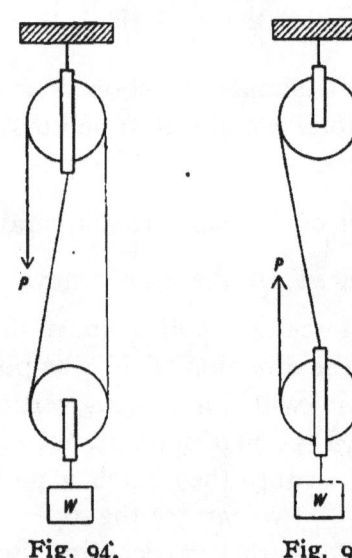

Fig. 94.   Fig. 95.

Combinations of Pulleys. There are two ways, illustrated in Figs. 94 and 95, by which a fixed pulley and a movable one may be used in combination. Later on it will be shown that these combinations are not essentially different, but for the present we will consider them separately.

In the first case only one end of the rope moves when the weight is raised, the other end being attached to the fixed pulley. In the other case every part of the rope moves when the machine is in operation.

In the first case the applied force would move twice as fast as the weight and hence $P = W/2$. For "gaining power" this arrangement of the two pulleys has no advantage over the single movable pulley, but it has the possible advantage that by passing the rope over the fixed pulley, the force is exerted downward instead of upward. This difference is shown in Figs. 96 and 97, placed side by side. If the rope were not passed over the upper pulley, as shown by the dotted line, this pulley would serve no purpose whatever; and even when it is called into use the free end of the rope does not move any faster for that reason, and

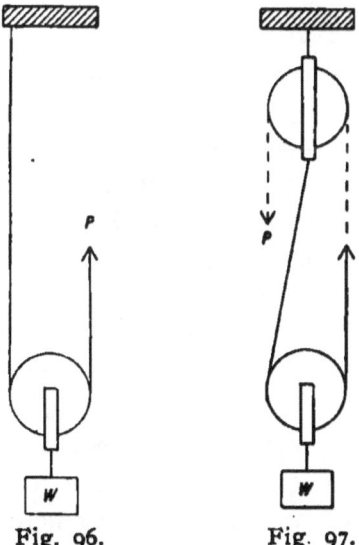

Fig. 96.   Fig. 97.

there is no further gain of power,—which is true, as we have learned, of *all* fixed pulleys. The weight drags down on two ropes, each of which supports $W/2$, and the free end of the rope merely balances one of these.

In the second case (Fig. 95), where one end of the rope is attached to the movable pulley, $P = W/3$, as may be readily proved by the principle of work. Statically, this follows from the fact that the weight is supported equally by the three ropes.

In these two cases the load sustained by the point of support is very different. In the first case it is $3/2\ W$; in the second case it is $2/3\ W$. Because, in the first case there are three ropes dragging downward and each is under a tension of $W/2$; while in the second case there are only two ropes, whose tensions, each $W/3$, exert a downward force upon the support. In each system if we replace the weight by a rigid support, the pull will be $3\ P$ on one support and $2\ P$ on the other. In other words, the second system is the same as the first, reversed end for end.

From the relations just deduced it will be observed, as a general principle of pulleys, that if we disregard friction, etc., the tension is the same in all parts of the rope*; and this tension is equal to the quotient of total load at either end of the system—either the weight or the load sustained at the point of support—divided by the number of parts of the rope at that end. For example, in the first of the two cases just considered, the load on the support was $3/2\ W$, in the same ratio as the number of ropes $3:2$; and the tension throughout the rope is $W \div 2$, or $3/2\ W \div 3$, the former being the quotient for one end of the system, and the latter being the quotient for the other end. In the second case the load on the supporting point was $2/3\ W$, the number of ropes at the two ends of the system being in the same ratio $2:3$, and the tension in the rope being $W/3$.

---

*This does not apply to the *different* ropes used in the systems illustrated in Figs. 99, 100 and 101.

From this consideration it follows, as we have already asserted, that there is no essential difference between these two cases; when a pull is exerted on the free end of the rope, a force is called into action at each end of the system of pulleys. We call the "movable pulley" the one that moves first. The fact that the other end did not move, or that the two ends did not move together, is a circumstance entirely independent of any consideration essential to the pulley. Since pulleys are used to overcome some resistance

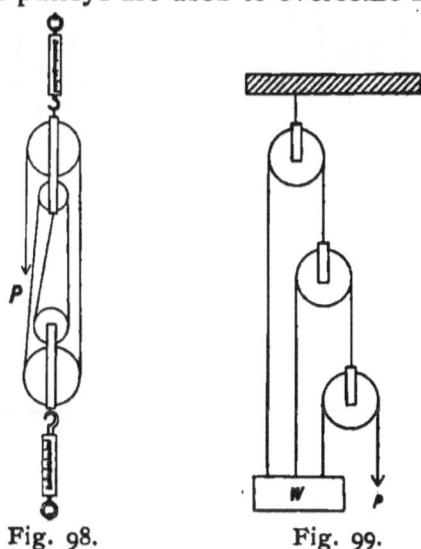

Fig. 98.   Fig. 99.

by fastening the other end of the system to a fixed support, the movable end is always the one at which the "weight" W, is assumed to be placed.

Exercises:

1. *A system of four pulleys is arranged in the manner shown in Fig. 98, and the two ends are connected with spring balances. If a pull of 2 lbs. is exerted at the free end of the rope, what will be the reading of each of the spring balances?*

Disregard friction, weight of pulleys, etc. In practice, the pulleys would be of the same size and would be placed side by side in each "block." They are shown differently in the diagram for the sake of clearness.

2. *When the upper balance reads 100, what is the magnitude of the applied force, and what is the reading of the other balance?*

3. *Find the relation between* P *and* W, *the tension in each rope, and the load on supporting beam, for the system of pulleys shown in Fig. 99.*

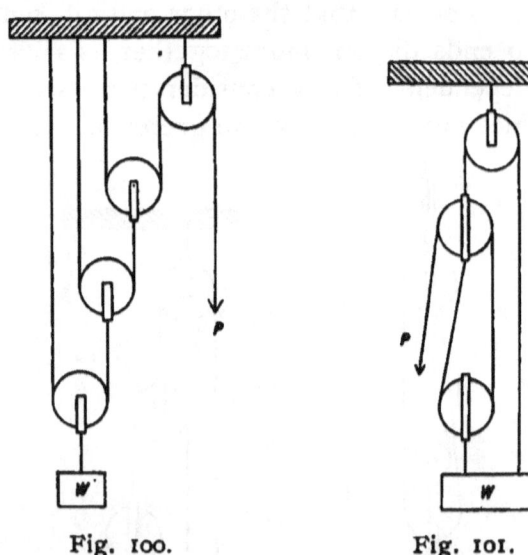

Fig. 100.   Fig. 101.

4. *Find the relation of* P *and* W, *the tension in each rope, etc., for the system shown in Fig. 100.*

5. *Find the relation between* P *and* W, *tensions in ropes, etc., for the system shown in Fig. 101.*

THE INCLINED PLANE. As applied to tools and machines the Inclined Plane is represented in chisels and other edge-tools, nails, screws, wedges, cams, eccentrics, propeller blades, etc. The name is derived from the primitive device of a plane surface used to elevate a "dead load" to some desired altitude by means of any "easy" incline—a skid, for example.

The steepness of a gradient or slope is sometimes designated by the number of degrees in the angle between the inclined plane and the horizon, but it is more commonly expressed as a percentage. This percentage is the quotient of $\frac{BC}{AB}$, (Fig. 102)—the altitude and base of a right triangle, of which the hypotenuse is any part of the inclined surface.

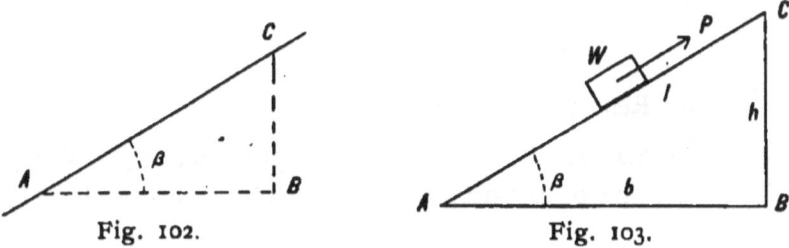

Fig. 102.   Fig. 103.

The mechanical advantage of the inclined plane depends upon the steepness of the gradient and the direction in which the applied force is impressed. Let the right triangle ABC (Fig. 103) represent an inclined plane, of which AB is the horizontal base. Let the base, altitude, and hypotenuse be represented by b, h and l, respectively. To prevent the weight W from sliding down the plane let a **force be applied in a direction parallel to the slope,** or hypotenuse, AC. If there were no friction what would be the relation between P and W to produce equilibrium? This problem can be solved by either of two methods: (1) by the principle of work, or (2) by resolving the gravitational action, W, into two components parallel and perpendicular to the plane.

1. **By the Principle of Work.**

Since gravity acts vertically, no work is done in moving a body horizontally with a uniform velocity, (unless there is friction). Hence, the work done *against* gravity in moving

W from A to C is Wh—the same as if it had been raised vertically from B to C. The work done *by* P is measured by the distance moved in the direction in which it is applied, and is equal to Pl.

By the principle of work $Pl = Wh$,

or
$$P = \frac{h}{l} W.$$

2. **By Resolution of the Vertical Force, W.**

If we resolve W into two components, w_1 and w_2, (Fig. 104), one perpendicular and the other parallel to AC,

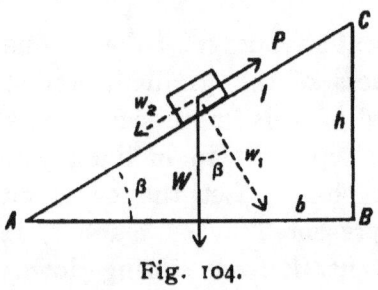

Fig. 104.

the former will have no tendency to move the weight one way or the other on the plane. The latter component, w_2, is the only part of W that P is required to equilibrate.

Hence, to produce equilibrium

$$P = w_2 = W \sin \beta.$$

From the triangle ABC,

$$\sin \beta = \frac{h}{l}, \quad \text{whence}$$

$$P = \frac{h}{l} W, \text{ as already proved.}$$

When P is not Parallel to the Plane. When P is applied in a direction making an angle γ with the hypotenuse, as shown in Figs. 105 and 106, it is only partially effective in the direction AC, the component p_1 perpendicular to AC being entirely useless in any direction at right angles to itself, (except to alter the amount of friction between W and the plane, which we are now disregarding). If p_2 is the component of P parallel to AC, then by the last paragraph

$$p_2 = \frac{h}{l} W = W \sin \beta$$

But $p_2 = P \cos \gamma$, whence $P \cos \gamma = W \sin \beta$, or $P = W \dfrac{\sin \beta}{\cos \gamma}$.

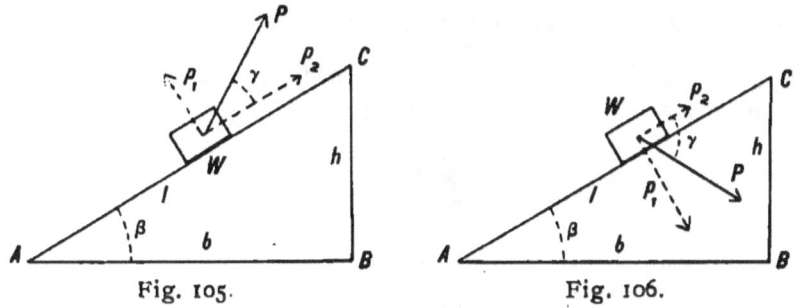

Fig. 105.   Fig. 106.

A very important case arises when the direction of P is parallel to AB. The angle γ is then equal to β. Using the expression

$$P = W \frac{\sin \beta}{\cos \gamma},$$

if $\gamma = \beta$, it becomes $P = W \dfrac{\sin \beta}{\cos \beta} = W \tan \beta$.

Referring to the triangle it will be seen that

$$\tan \beta = \frac{h}{b}, \text{ whence } P = \frac{h}{b} W.$$

That is, if the power is applied parallel to the base, P and W are in the inverse ratio of the base and height. This, of course, is verified by the principle of work; for when the weight is moved up the plane from A to C it moves through the horizontal distance b, and the vertical distance h; and P being applied horizontally does an amount of work equal to Pb, which we know must be equal to Wh. If $Pb = Wh$, then

$$P = \frac{h}{b} W$$

The Wedge. It has just been shown that when the "weight" on an inclined plane acts in a direction perpendicular to the base, while the power is applied parallel to the base, the relation between P and W is $P = \frac{h}{b} W$. A practical illustration of this would be the device shown in Fig. 107. A rod fitted into the guides G and G_1, and

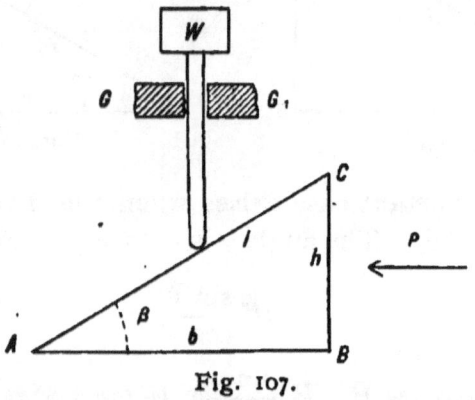

Fig. 107.

hence free to move only in a vertical direction, is held down against an inclined plane by means of a weight W. A pressure P is applied to the back of the inclined plane, which is pushed, wedge-like, under the rod. If the first position is such that point A is vertically under the rod, and the plane is pushed through its entire length, until C comes under the rod, the latter will have been raised through a vertical distance BC, while P moves through the horizontal distance BA. By the principle of work $Pb = Wh$, or $P = \dfrac{h}{b} W$.

It happens, however, in devices where the wedge is used, that W is not always, or even usually, vertical,—that is, perpendicular to the base of the plane. Very frequently it is perpendicular to the hypotenuse, as in the case of a carpenter's chisel used to split a piece of wood in the manner shown in Fig. 108. If the angle at the edge of

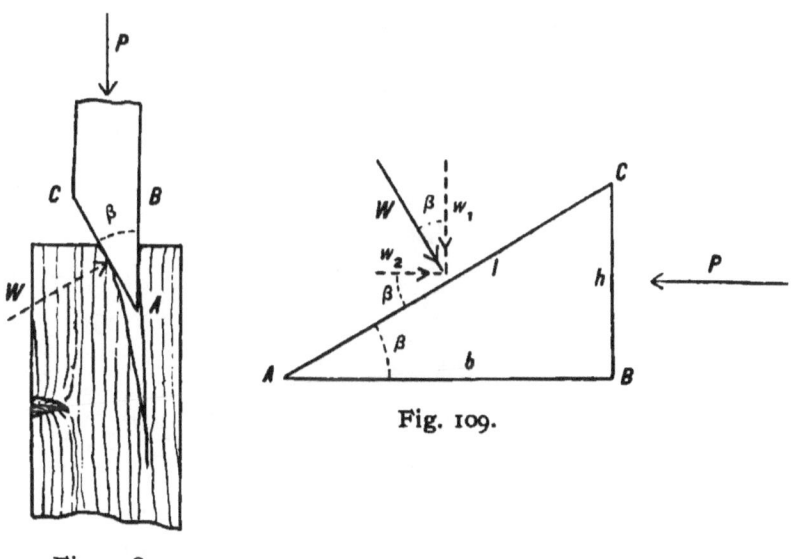

Fig. 108.

Fig. 109.

the chisel is β, and the cohesion of the wood causes a pressure W at right angles to the slope, then the conditions of the problem are as represented in Fig. 109. W is resolved into components perpendicular and parallel to AB, and the problem is solved by statics or by the principle of work.

Statically, the component w_2 is the only part of W opposing P and tending to prevent the wedge of the chisel from pushing into the wood, (for the component w_1 can exert no influence in a direction at right angles to itself). And $w_2 = W \sin \beta$. To produce equilibrium $P = w_2$, or $P = W \sin \beta$.

Or, by the principle of work, if the wedge of the chisel is pushed into the wood from A to C, the spreading of the wood through a distance BC will signify that the component w_1 is overcome through that distance, requiring an amount of work, or expenditure of energy, equal to $w_1 \times h$. (The component w_2 is not now considered, for the

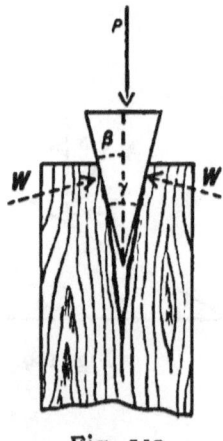

Fig. 110

reason that its point of application is not moved through any distance in its own direction,—that is, parallel to AB). Since the force P, parallel to the base, does this work by moving through the distance BA, or b, we have $Pb = w_1 h$,

or $$P = \frac{h}{b} w_1.$$

But $\frac{h}{b} = \tan \beta$, and $w_1 = W \cos \beta$.

MACHINES.

Whence by substitution, $P = W \sin \beta$, as already proved by the static method.

In most cutting tools the edge has the V form of the typical wedge, illustrated in Fig. 110. A wedge of this sort is obviously equivalent to two inclined planes, having a common base—as represented by the vertical dotted line. By driving the wedge full length into the wood, the fibres are spread twice as far as if the wedge had been a right triangle instead of isosceles. The resistance W is taken as being perpendicular to each face. Therefore, the relation between P and W is $P = 2 W \sin \beta$. The same relation is sometimes expressed in terms of the total angle at the vertex of the wedge, or $P = 2 W \sin \dfrac{\gamma}{2}$.

In the ordinary use of a wedge for rough work—such, for instance, as we have selected for our illustration—the amount of friction is very great. But in cases where accurate calculations would be at all necessary the friction would be reduced to a reasonable limit, where it could be taken into consideration and properly allowed for in the computations.

The results of the preceding computations for the Inclined Plane and Wedge are recapitulated in the following table, showing the relations between P and W for the different directions in which these forces are usually applied :

1. $\begin{cases} W \text{ perpendicular to base} \\ P \text{ parallel to hypotenuse} \end{cases}$ $P = \dfrac{h}{l} W = W \sin \beta$

2. $\begin{cases} W \text{ perpendicular to base} \\ P \text{ parallel to base} \end{cases}$ $P = \dfrac{h}{b} W = W \tan \beta$

2. $\begin{cases} W \text{ perpendicular to hypotenuse} \\ P \text{ parallel to base} \end{cases}$ $P = \dfrac{h}{l} W = W \sin \beta$

3. $\begin{cases} \text{For "}V\text{" wedge} \\ W \text{ perpendicular to sides} \\ P \text{ perpendicular to back} \end{cases}$ $P = 2 \dfrac{h}{l} W = 2 W \sin \beta$

The Screw. The Screw is a combination of the Inclined Plane (in a modified form) and the Wheel-and-Axle. The wheel-and-axle element is illustrated by taking a forged bolt before the thread is cut and applying a wrench to the head. The wrench will turn in a large circle, constituting the "wheel," and the shank of the bolt will be the "axle." Cutting the thread adds the element of the inclined plane, as may be shown by the following simple experiment. Cut a piece of paper into triangular form, to illustrate an inclined plane. The grade or slope of this plane will depend upon the diameter of the bolt and the desired pitch of the screw. Place the edge BC of the plane against the side of the bolt and parallel to the axis of the cylinder. By wrapping the paper continuously around the bolt, the upper edge CA, representing the inclined surface, will describe a helix, which will coincide with the path along which a thread could be cut of the desired pitch.

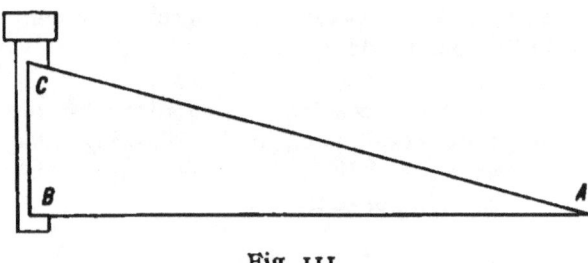

Fig. 111.

The pitch of a screw could be defined in the same manner as an ordinary inclined plane—as an angle measured in degrees, or as a percentage—but it is easier to make computations directly from the number of threads per inch, measured parallel to the axis. A machine screw is measured by two numbers (besides the length), one referring to the diameter of the shank and the other designating the number of threads per inch. A number 16 pitch means 16 threads per inch.

The form of a thread is sometimes rectangular, sometimes V-shaped, and sometimes a modified form with rounded edge or vertex. But the shape or angle of the thread does not alter the angle of pitch.

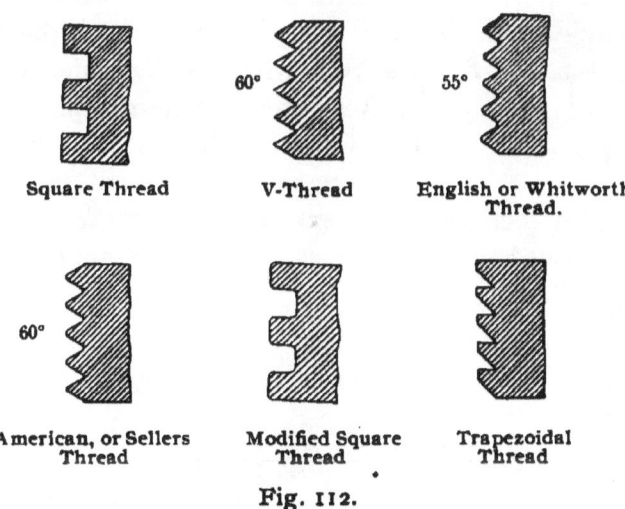

Square Thread   V-Thread   English or Whitworth Thread.

American, or Sellers Thread   Modified Square Thread   Trapezoidal Thread

Fig. 112.

The operation of a machine screw or a bolt requires a nut. Both bolt and nut may be movable, or one of them may be fixed,—according to needs. A wood screw is always used in soft material, which is compressed into necessary grooves by the thread of the advancing screw.

When a force is properly applied to the screw, the action of the thread is analogous to that case of the inclined plane where P is applied to the back of the plane and W acts perpendicular to the base. There is the additional consideration of course, of the wheel-and-axle. It is not necessary, however, to follow out all these relations, because the total mechanical advantage of a screw can be computed directly from its pitch by the principle of work. Suppose, for instance, that a bolt of ½-inch diameter, and number 8 pitch, is turned by means of a wrench applied to the head. The resistance W, let us assume, is occasioned by compressing two pieces of wood held together between the nut and the head of the bolt. If a pressure P is exerted on the wrench at a point 9 inches from the axis of the bolt, what total resistance W can it overcome? One complete turn of the wrench will cause the bolt to advance through the nut ⅛ inch (the distance between two adjacent threads), or $\frac{1}{96}$ foot. The distance moved by the point at which P is applied will be

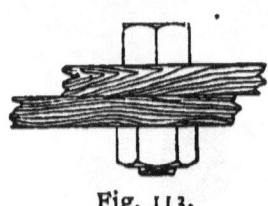

Fig. 113.

$$\frac{2\pi \times 9}{12} \text{ feet.}$$

Therefore

$$P \times \frac{2\pi \times 9}{12} = W \times \frac{1}{8} \times \frac{1}{12}.$$

In general, if h is the distance between two adjacent threads, and r is the distance from the axis of the screw to the point of application of P, the relation between P and W would be expressed by the formula

$$P \times 2\pi r = Wh.$$

The two distances r and h must be expressed in the same units,—both in inches or both in feet. In the problem taken as an illustration these distances were reduced to feet in order to get the foot-pound as a unit of work.

MACHINES. 155

Notice that W is supposed to act parallel to the axis of the screw,—which would be perpendicular to the base of the inclined plane.

Notice, also, that the diameter of the bolt is not used in the calculations. It is duly involved, however, in the pitch, but was eliminated from the computations when we agreed to deduce the mechanical advantage directly from the pitch. The pitch, h, is equal to $\pi \times d \times \tan \beta$, where d is the diameter of the bolt and β is the slope of the thread measured in degrees. Each thread is equivalent to an inclined plane of height, h; base, πd; and slope β, wrapped around the bolt.

Endless Screw. Consider a threaded cylinder fitted into guides in the manner shown in Fig. 114. The screw itself cannot advance, being restrained by the guides, but the

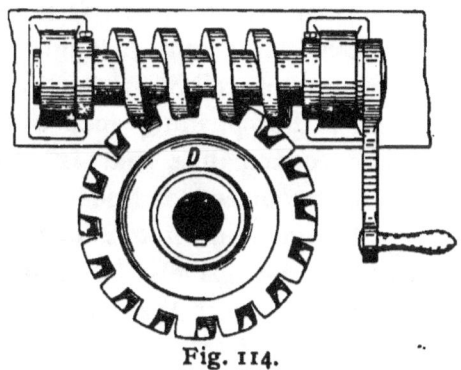

Fig. 114.

toothed wheel D, can be made to rotate continuously by turning the screw. The teeth on the circumference of the wheel feed into the threads of the screw from one side and out at the other. From the wheel D power may be taken, after the manner of the wheel-and-axle, and the action is endless, notwithstanding the limited length of the screw.

The Cam and Eccentric. If a circular disc is pivoted at a point other than the center, as the point O (Fig. 115), and is made to rotate around that point it is said to have an eccentric motion, and is itself called an eccentric. If a rod were set in guides so as to rest vertically upon point S, it would be pushed up vertically from S to T' by a semi-rotation of the eccentric. If the eccentric is made to rotate continuously, the rod will move up and down with a reciprocating motion between positions S and T'.*

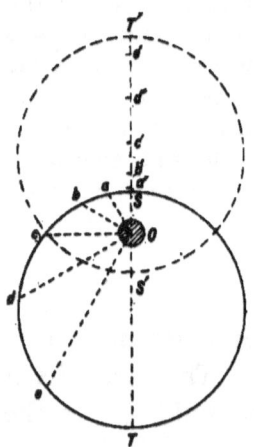

Fig. 115.

The total work done in raising the rod from S to T' is independent of the manner in which it is accomplished, but it is not proportioned uniformly throughout the stroke. It is equivalent to pushing a weight up an inclined surface of varying slope,—or rather pushing such a surface under the weight. Suppose, for instance, that six equal angles be constructed around O by means of lines Oa, Ob, Oc, etc. As rotation takes place, the points a, b, c, d, e and T assume positions a', b', c', etc., successively. During the first $\frac{1}{12}$ rotation the rod is raised through the distance Sa', equal to the difference between Oa and OS; during the next interval of $\frac{1}{12}$ rotation the rod is raised from a' to b', the difference between Ob and Oa, etc. Since $a'b'$ is greater than Sa', it is

---

*The eccentric is used upon engines to produce and control the valve motion, but the motion thus given to the eccentric rod is not the same as that given to the rod mentioned in the above illustration. A valve eccentric, running in an eccentric-strap, gives a motion that is the same as if a crank of the same swing had been used instead of the eccentric.

obvious that the average slope of the surface pushed under the rod during the second interval was greater than during the first interval.

The distance ST' is called the "swing" or "throw" of the eccentric.

A cam is essentially the same as an eccentric, differing mainly in the shape of the periphery. By varying the surface of the disc an infinite number of straight-line motions can be accomplished. By means of a certain heart-shaped disc (Fig. 116) the rod can be given a uniform velocity. If the disc is not symmetrical the return stroke will be different

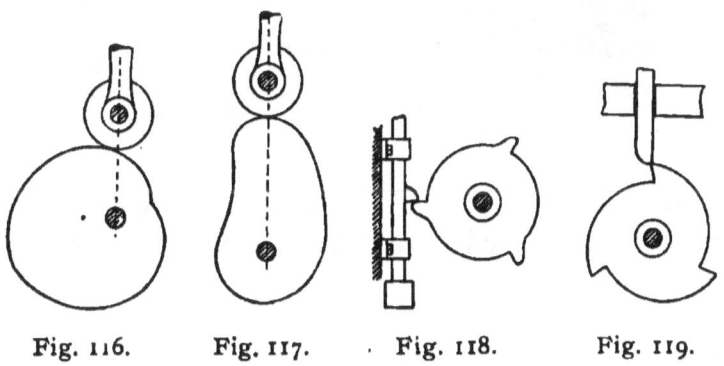

Fig. 116.   Fig. 117.   Fig. 118.   Fig. 119.

from the out stroke (Fig. 117). Several repetitions of the straight line motion can be accomplished in a single rotation of the axis, by means of lugs or shoulders on the edge of the disc (Figs. 118 and 119). Sometimes the end of the rod plays in an irregularly grooved collar (Fig. 120). All these cases involve the Inclined Plane and can be solved by the principle of work if the swing is known.

The end of the rod that bears on the cam is usually fitted with a roller, for which allowance must be made in constructing the cam to give a required motion to the rod. Very frequently the rod is connected with a system of levers to still further modify the motion (Fig. 121). By the

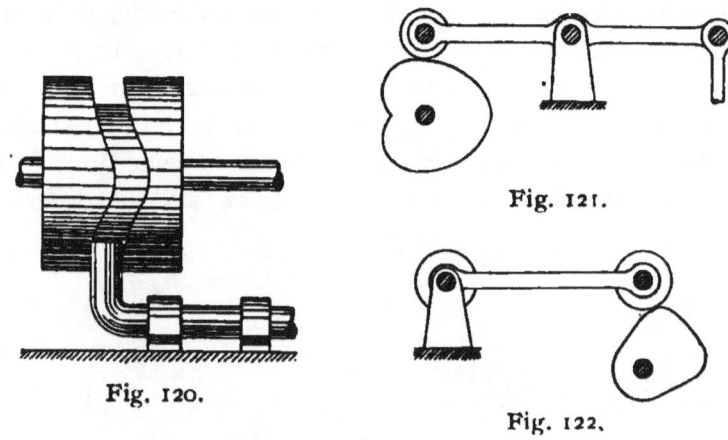

Fig. 121.

Fig. 120.

Fig. 122.

arrangement shown in Fig. 122, uniform motion of the shaft bearing the eccentric is converted into irregular motion of a second shaft.

The Toggle-Joint. The Toggle-Joint is used in adjustable carriage tops; in printing presses; in machines used for stamping metals, leather, wood, etc.; and occasionally in copying presses when the desired pressure is greater than could be obtained from a screw. Its mechanical advantage is enormous.

Two bars, AB and AC are pivoted in the manner shown in Fig. 123, and the ends B and C are constrained to move laterally, or at right angles to the path of A. A small pressure at A, applied as shown, will produce a very great outward pressure at B and C, depending upon the magnitude of the angle β.

The relation between P and W is not fixed, but changes as the angle of the joint changes, the mechanical advantage of P over W becoming greater and greater as the rods flatten out. If a constant force P is applied at A it will give a certain greater pressure, W, at B and C, but as the joint gradually straightens out, approaching the straight line BC, the value of W becomes greater and greater, though the applied force P has remained constant. As point A moves through the distance AD, the end B moves through

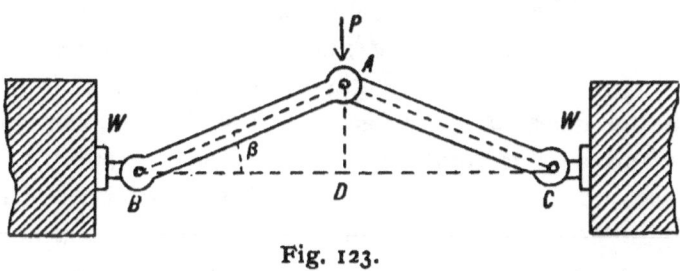

Fig. 123.

the difference between AB and DB. Now $DB = AB \cos \beta$, and for small angles $\cos \beta$ is almost at its maximum value,— that is, very nearly equal to one,—whence the difference between AB and AD is very small. Even when β is as large as 5 degrees, we find $\cos \beta = 0.9962$, and $AB - BD = .0038\ AB$. As BAC approaches a horizontal position as a limit this difference becomes almost infinitesimal in comparison with an appreciable movement of P in direction AD, and therefore, at this instant—just before β becomes zero, according to the principle of work, W is enormously greater than P.

Differential Motion. In the differential pulley, the differential screw, and the differential wheel-and-axle, an extra part is added to the machine with the sole function of reducing the motion of the load and thereby increasing the mechanical advantage of the machine.

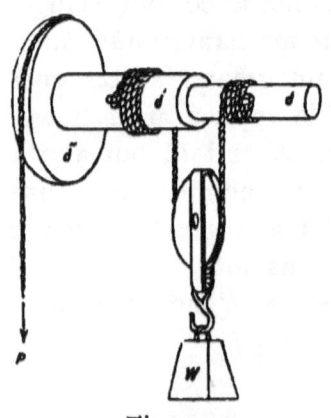

Fig. 124.

In Fig. 124, illustrating a differential wheel-and-axle, P is applied to the circumference of a wheel of diameter d''. The weight is suspended from a pulley which is itself supported by ropes coiled around the cylinders d and d', constituting two portions of a solid or continuous axis, upon which the wheel d'' is fixed. If the force P causes the wheel and axis to rotate, the rope wrapped around d' will be drawn up, but at the same time the portion coiled upon the cylinder d will be unwound. The rate at which W will be raised will depend upon the difference between the circumferences of the two parts of the axis (though we must remember also that the pulley supporting the weight is a movable pulley, for which reason the weight rises only half as fast as the rope is shortened).

Everything duly considered, it will be found that

$$P \times \pi d'' = W \times \left(\frac{\pi d' - \pi d}{2}\right), \text{ or}$$

$$P \times d'' = W \left(\frac{d' - d}{2}\right);$$

$$P = W \frac{d' - d}{2 d''}.$$

The chain-hoist (or differential pulley, as it is called), and the differential screw are perhaps the most familiar devices in which the idea of differential motion is used. In the differential pulley, two wheels of slightly different diameters are fixed on the same axis, after the manner of a wheel-and-axle. A third wheel supports the weight in the manner shown in Fig. 125. Instead of a rope, as ordinarily used with pulleys, an endless chain passes from pulley d_1 to the movable pulley (part h); thence to the pulley d (part k); thence downward, hanging free (part l); and thence to the circumference of d_1 (part m), thus completing the circuit.

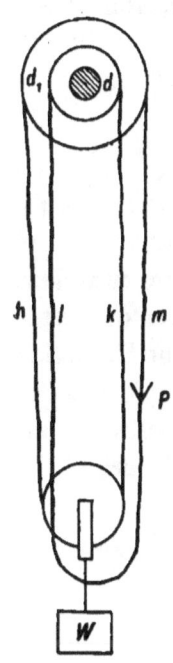

Fig. 125.

Now if P is applied to part m of the chain, the fixed pulleys d and d_1 will rotate together; part h of the chain will move upward as far as P moves downward, and at the same time part k is fed downward from the circumference of d. But since d and d_1 are fixed to each other and rotate together, the part h is taken up faster than part k is fed downward, and hence in a single rotation of these two wheels W will be raised a distance equal to *one-half* the difference between their circumferences and P will move downward through a distance equal to the circumference of the larger. Therefore

$$P \times \pi d_1 = W \times \left(\frac{\pi d_1 - \pi d}{2}\right), \text{ or}$$

$$P \times d_1 = W \left(\frac{d_1 - d}{2}\right), \text{ or}$$

$$P = W \frac{d_1 - d}{2 d_1}.$$

The two wheels or cylinders that contribute to the differential motion can be made so slightly different that the motion of W may be made as slow as we please, and thus a very small applied force can be made to lift an enormous weight, up to the limit of the strength of the machine.

Compound Machines. Many hand tools represent only one of the different elementary machines enumerated on p. 111, but most mechanical contrivances are more complex. A compound machine is one made up of a number of simple machines. The "Weight" of the first becomes the "Applied Force" for the next, as in compound levers (p. 117), and the train of wheels (p. 136),—though it is not necessary that the successive parts of the machines should all be of the same kind. In moving a house on rollers, for example, a capstan is used in combination with a compound system of pulleys.

Examples:

1. If two forces of 60 pounds each, acting on the same point at an angle of 60° with each other, are exactly balanced by a third force making an angle of 120° with the two given forces, find the magnitude of the third force.

2. A straight uniform lever AB, *12 feet long, balances about a point in it 5 feet from* B, *when weights of 9 pounds and 13 pounds are suspended at* A *and* B, *respectively. Find the weight of the lever.*

3. An iron bar 6 feet long is free to turn about a horizontal axis which is 4 feet from one end. On top of it is placed a second bar 4 feet long, but otherwise like the first bar, so that the lengths of the bars are parallel. The two bars are now balanced upon the pivot. Describe the position of the top bar.

4. *Analyze the action of a claw-hammer in drawing a nail. Where is the fulcrum? The "weight"? The "power"?*

5. *Why does the driver of a heavy load take a zig-zag course in climbing a hill?*

6. *The drum of a windlass has a diameter of 10 inches, and the crank has a radius of 18 inches. What minimum force must be applied to the handle to raise a load of 160 pounds?*

Disregard friction and thickness of rope.

(b) What difference if we allow a half inch diameter of rope?

7. *The screw of a letter-press has a pitch of ¼ inch, and the diameter of the wheel is 10 inches. What pressure will be exerted upon the copying book by forces of 25 pounds applied at two different points on the circumference of the wheel?*

8. A two-arm balance is supposed to have arms of equal length; otherwise it is false. A false balance may appear to be true because the beam stands at equilibrium when there is no load in the pan, but a little thought will show that it will not continue at equilibrium when equal weights are added.

(a) If a dishonest dealer were using a false balance of this sort, in which arm would he place the substance being sold?

(b) Can you think of any way by which to test a balance to determine whether it is true or false?

9. *In what way do metal-cutting shears differ most from ordinary scissors? Why?*

10. *In the chain-hoist described on p. 161, the two wheels d and d_1 have diameters of 6½ inches and 7 inches, respectively. What force P must be applied for a load W equal to 600 pounds?*

11. *A 1200-pound anchor is hoisted by means of a capstan having a drum of 18 inches diameter and four spokes each 8 feet long.*

What force must be exerted by each of four sailors pushing at the extreme ends of the spokes?

12. A plank lies with one end projecting over a log. A boy weighing 100 pounds walks out on the projecting end, and when he gets 6 feet from the log, the plank tips. The center of gravity of the plank is 5 feet from the log. Find the weight of the plank.

13. A plank weighing 200 pounds lies with one end projecting over a log. A boy weighing 100 pounds walks out on the projecting end, and when he gets 6 feet from the log the plank tips. Where is the center of gravity of the plank?

14. A house on rollers is moved by means of pulleys and a capstan.

(a) If the resistance to rolling is 20 tons, at what rate can it be moved by a single horse working at the rate of one H. P.?

(b) If the drum of the windlass has a diameter of 20 inches and the system of pulleys has two wheels in each block, what force must the horse exert at a point on the arm of the capstan 14 feet from the center of the drum? And at what rate must the horse walk in order to move the house at the rate computed in part (a)?

15. For the endless screw shown on p. 155, assume the following dimensions:
> Pitch of screw, 1 inch.
> Crank arm, 12 inches.
> Number of teeth in wheel D, 16.
> Diameter of shaft bearing wheel, 1½ inches.

What force must be applied to the crank handle to lift a load of 500 pounds applied to the perimeter of the shaft?

CHAPTER VI.

FRICTION.

When one body moves in contact with another, whether by sliding or by rolling, or when an object travels through a fluid medium, as a bullet through the air, we instinctively assume that a frictional resistance inevitably accompanies the motion. The laws of friction are simple in statement, but they are quite empirical and do not always hold with that rigid mathematical accuracy which characterizes other laws of mechanics. This is because the usual phenomenon of friction, simple as it seems, is really an aggregation of a great number of less obvious phenomena, or of several groups of such aggregations, which taken together give rise to a complexity of conditions surpassing the possibility of mathematical analysis. For example, when one surface slides over another the frictional resistance is supposed to be occasioned mainly by the interlocking of an infinite number of particles on the two adjacent surfaces. No surfaces are smooth enough not to permit of friction; two "perfectly smooth surfaces" (if such were possible) when placed in contact with each other, and the air excluded, would be close enough together for complete inter-molecular action between the two surfaces, and the two bodies would cohere if of the same material, (or adhere if of different materials), so firmly that they would be as a single rigid body. No doubt the friction between two ordinarily smooth surfaces is caused in part by molecular attractions as well as by the interlocking of physical particles, or particles having appreciable dimensions.

Still more complicated is the frictional resistance between lubricated surfaces, or of a moving carriage. The rolling friction of the tires in contact with the ground, the sliding friction in the bearings, and the frictional resistance of the air, (each in itself a complex phenomenon), are all included in the total tractional resistance of the carriage.

It will be convenient to consider in order the laws, (*a*) of Sliding Friction; (*b*) of Rolling Friction; (*c*) of the Friction of Ropes and Belts on Pulleys, and (*d*) the Use of Lubricants.

Sliding Friction. For the purpose of illustrating the laws of sliding friction, the following results obtained by actual experiment may be used:

An ordinary building brick, smoothed on one face by means of a grindstone, was placed on a planed cedar board. A string passed around the brick was attached to a spring balance. By pulling continuously on the balance it gradually reached a tension at which the brick began to slide. Restoring the brick to its first position under the same conditions as before, it was found to move again at about the same reading of tension on the balance,—2.5 pounds.

A block of Oregon pine, of quite different dimensions from the brick, was placed on the cedar board as nearly as possible under the same conditions as the brick, and was found to slide when the pull was only 1.1 pounds.

How shall we account for this difference—2.5 pounds to slide the brick and only 1.1 pounds to slide the pine block? Was it due to the difference in the weights of the two; or to the area of the surface of contact; or to a difference in the natures of their surfaces; or to all these causes in varying degrees?

To answer these questions the observations were continued as follows:

(1) A second pine block of exactly the same dimensions as the brick was loaded with weights until it had also the same weight as the brick, 5.6 pounds. Placed on the cedar board under the same conditions as before, it was found to slide under a pull of 3.8 pounds.

Now, since the brick and the block have the same weight and the same area of surface in contact with the board, the difference between the friction of the brick and cedar (2.5 pounds) and the friction of the pine block and cedar (3.8 pounds) must be due to the fact that the nature and condition of the brick surface is quite different from the pine block.

In general, the friction between two surfaces tending to prevent sliding is different for different pairs of substances. It has been found in practice that **friction is generally greater between surfaces of the same kind**, as between steel and steel; leather and leather; etc. Hence the advantage of using brass bearings for steel shafts, and of covering with leather the face of a pulley used with leather belting.

(2) The opposite side of the brick had not been ground. Placing the brick on the cedar board with the rough side downward, the friction was found to be 2.8 pounds as against 2.5 for the smoothed surface.

The sliding **friction depends upon the roughness** of either or both **of the surfaces** in contact. It is obvious however, that this roughness cannot be measured mathematically.

(3) One of the long, narrow faces of the brick had been ground and the opposite face left unground. Placed with the smooth edge on the cedar board the friction was found

to be about 2.5 pounds, the same as for the broad side similarly ground. The friction for the ground edge was the same as for the rough side of greater area.

In general, the **friction between two surfaces is independent of the area of contact** This law is true within broad limits, but when one of the bearing surfaces is so small that it digs or cuts into the other, or when the pressure is so great that the surfaces are deformed or abrased, the law is interfered with by new considerations, quite distinct from frictional influences.

The most important law—that the **friction between two surfaces is proportional to the force pressing them together**—is deduced from the next observations.

(4) The Oregon pine block used above was loaded so as to weigh 4 pounds, and again placed on the cedar board. The friction determined in the same manner as before, was found to be 2.6 pounds. By placing additional weights on the block, equal to a total of 8 pounds, the friction was increased to 5.25 pounds, or about twice as much as before. In the same manner, when the weight or pressure was made three times as large (12 pounds) the friction was found to have been increased in almost the same ratio—to 7.9, according to the actual measurement.

That is, for two given surfaces, the friction depends only upon the pressure between them. If F is the friction, and P is the force pressing the surfaces together, then $\dfrac{F}{P}$ is constant.

The ratio $\dfrac{F}{P}$ is called the **Coefficient of Friction** for the given surfaces. This idea is used so much in practice that the coefficient of friction is usually represented in formulæ by

FRICTION.

a special symbol,—generally the Greek letter ϕ. For example, according to our measurements the coefficient of friction between the given surface of Oregon pine and cedar as used in the experiment, would be

$$\phi = \frac{3.8}{5.6} = .68 \text{ for the first measurement.}$$

$$\phi = \frac{2.6}{4} = .65 \text{ for the second measurement.}$$

$$\phi = \frac{5.25}{8} = .66 \text{ for the third measurement.}$$

$$\phi = \frac{7.9}{12} = .66 \text{ for the fourth measurement.}$$

Average $\phi = .66$

When we say that the coefficient of friction between the given surfaces of pine and cedar is .66 we mean that the total friction between them is always that fraction of the total pressure.

If the pressure acts in a direction not perpendicular to the two surfaces, as by pressing a stick obliquely against the block after the manner shown in Fig. 126, then it becomes necessary to resolve the pressure into two components. Only that part which acts in a direction perpendicular to the surface of contact is taken into account in considering the friction; the component parallel to the surface of contact will tend to push the block along the plane, but it does not thereby change the friction one way or the other.

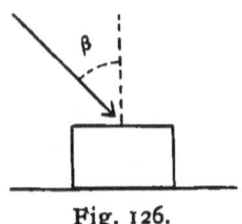

Fig. 126.

Example:

In the following table, containing the values found in the preceding experiments, insert in the column headed "ϕ" the value of the coefficient of friction for each pair of surfaces named in the same horizontal line.

SURFACES		P	F	ϕ
I	II			
Brick (On ground face)	Cedar (planed)	5.6	2.5	
Oregon Pine (Planed; with weights added)	" "	5.6	3.8	
Brick (On ground edge)	" "	5.6	2.5	
Cast Iron (Dressed on shaper)	" "	8.7	4.5	
" "	Leather (Old belting; flesh side)	8.7	1.8	
" "	Leather (Same piece; hair side)	8.7	1.5	
Oregon Pine	Cedar (planed)	4	2.6	
" "	" "	8	5.25	
" "	" "	12	7.9	

In all our illustrations, for the sake of simplicity, we have considered the pressure between the surfaces as being due to the weight of an object. But **the laws of friction apply to pressures of all kinds**, as in a brake operated by levers or by a spring. As a further example, in dealing with the friction in the bearings of a line of a shafting we have to consider not only the downward pressure due to the dead weight of the shafting and pulleys, but also the force exerted by the tension of the belt (which may be in any direction); it is the resultant of the two that gives the total pressure of the shaft in the bearings—that pressure from which we must compute the friction.

The use of sliding blocks for our purposes of illustration may also tend to still another misunderstanding. We are not dragging the weight of the block when we slide it on a *horizontal* surface—at least not in the sense that we overcome the weight in lifting it. The effort of dragging is not due to the weight of the block, except as the weight causes friction at the surface of contact. It should be remembered that if there were no friction and the plane were *perfectly* horizontal the slightest force would cause the block to move and to keep on faster and faster; of course, the greater the mass of the block and the smaller the force applied to it, the less rapidly would it gain velocity, but, as already stated on p. 75, if there were no friction, any force, howsoever small, would produce motion in any mass, howsoever large, on a horizontal plane.

Static Friction and Kinetic Friction. In any of the preceding observations, if we had watched the reading of the spring balance after the body had commenced to slip, we would have seen that the pull necessary to keep the body moving is less than just before slipping commenced. In other words, the coefficient of friction between the two surfaces when they are in relative motion is less than the friction of rest, or the statical coefficient. However, the friction during motion, or kinetic friction, obeys the same laws that we have already deduced for statical friction, the friction being simply less by a certain amount for the given surfaces, but still proportional to the pressure. This is true provided the velocity is not too great; at high speeds the coefficient is less, while at very low speeds the statical coefficient and the kinetic coefficient are nearly equal.

Not only is the statical greater than the kinetic friction, but it is also true that the longer the surfaces have remained in quiet contact the greater the statical coefficient.

Friction Always a Resistance. In discussing the idea of a force (p. 60), it was stated that friction is not an active agent capable of producing motion; in fact it has no existence until called into play by our effort to make the two surfaces slide in contact with each other. Furthermore, it has no direction of application such as a force always has; when a brick rests on a horizontal surface the friction is the same, whatever the direction in which we slide the body.

Determination of the Coefficient of Friction by Means of the Limiting Angle, or Angle of Repose. This simple method of measuring the coefficient of friction is exemplified in the following experiment:

A block of iron weighing 8.7 pounds was placed on the cedar board and the latter gradually inclined until the block

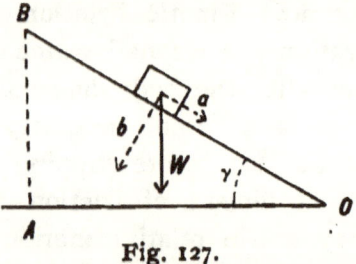

Fig. 127.

began to slip down the grade. The angle of elevation, γ, at which the slipping commenced was measured by applying an ordinary two-foot square in such manner as to determine the lengths OA and AB in Fig. 127. These lengths OA and OB were found to be 20 inches and $10\tfrac{7}{8}$ inches respectively. The ratio $\frac{AB}{OA}$ is simply tan γ, which, in this case is tan $\gamma = \frac{AB}{OA} = \frac{10.56}{20} = 0.53$. By comparing this value, 0.53, with the coefficient, 0.52, for the same surfaces (as previously determined by means of the spring balance), it will appear that the coefficient of friction is simply the tangent of the angle at which slipping commenced.

That this is true can be readily shown by simple geometric demonstration. The weight W is a force acting vertically downward, but as the plane of the board is no longer horizontal, only a part of W—the component perpendicular to OB—serves to press the two surfaces together. Accordingly, if W is resolved into two components, one parallel to OB and the other perpendicular to OB, the former (component a) tends to drag the weight down the plane, but cuts no figure in the friction, while the other, b, tends only to cause friction. If φ is the coefficient of friction, the total friction is $b\phi$, and if the component a is just sufficient for the body to slip down the plane, then obviously $a = b\phi$. But $a = W \sin \gamma$ and $b = W \cos \gamma$; whence $W \sin \gamma = \phi W \cos \gamma$, or

$$\phi = \frac{W \sin \gamma}{W \cos \gamma} = \tan \gamma, \text{ as was to be shown.}$$

Examples:

  1. *A body resting on a surface just begins to slide when the surface is inclined 21° 17'. What is the coefficient of friction?*

  2. *A body placed on a 40% grade has just sufficient inclination to cause slipping. What is the coefficient of friction?*

  3. *The coefficient between the Oregon pine and the cedar board was 0.66. What is the greatest angle at which the board could be inclined without the block slipping?*

  4. *If a body is on an inclined plane the total friction is less than on a horizontal plane. Why?*

Work Done in Dragging a Body by Sliding. When a body is dragged with a uniform velocity along a horizontal surface, the only work done is in overcoming the frictional resistance (See p. 75). If the body is on an inclined plane the total friction is less than on a horizontal plane, but there is the additional consideration that extra work will be required if the body is being moved up the incline. If the body is being moved down the grade a part of the weight of the body then aids in overcoming the friction.

Examples:

1. *Resolving the weight into two components, a and b, parallel and perpendicular to the plane (as in Fig. 127), and calling the coefficient of friction ϕ, what will represent the force necessary to move the body up the plane? What to move the body down the plane? In the latter question what is signified if a is greater than $b\phi$? If $a = b\phi$ what is the value of γ?*

2. *A mass weighing 200 pounds rests on a horizontal surface. If the coefficient of friction is 0.23 what force will be required to drag the body along the surface with a uniform velocity? What horse-power is required to drag this body at the rate of 100 yards a minute?*

3. *A body weighing 90 lbs. rests on a surface inclined at an angle $\gamma = 14° 3'$. The coefficient of friction between the two surfaces is 0.375.*

   (a) *What force is necessary to drag the body up the plane?*

   (b) *What force is necessary to drag it down the plane?*

   (c) *What horse-power will be required in each case to keep the weight moving at the rate of 40 feet per second?*

Rolling Friction. It has already been stated (Kinematics p. 32) that when a wheel rolls along the ground that part of the wheel in contact with the ground is always at rest; as each point comes down and touches the ground that point on the wheel is for the instant at rest relatively to the surface upon which the wheel rolls. There is no sliding of the wheel bodily along the ground, and yet there is friction, which in time will bring the wheel to rest.

It is this resistance that is called **rolling friction.** As the successive parts of the circumference come down to the ground it is not difficult to conceive how the minute projecting particles of the wheel collide with similar elements of roughness on the ground, producing this resistance. The rougher the surface the greater the rolling friction, as we know from the fact that a ball will roll farther on a smooth floor than on a carpet, or on a gravel path. If the surface of the wheel is soft it will be flattened and its progress retarded. Likewise, if the surface upon which a rolling body moves is such that a depression is formed by the weight of the body, a corresponding ridge will also be formed in the path of the object, so that the energy of the rolling body will be used up in deforming the surface and constantly climbing this little ridge.

If a wheel or ball rolls in contact with a curved surface, the "normal pressure" at any instant would refer to a direction perpendicular to a line or plane tangent in common to the two sufaces at their point of contact for the given instant.

Work Done in Dragging Vehicles. It has already been stated that when a rolling vehicle is moved, the tractional resistance includes all friction, whether sliding or rolling, both in the bearings and at the rims of the wheels. This resistance is designated as so many pounds to each ton of load, including the weight of the vehicle. For example, if a car weighing 12 tons carries a freight load of 6 tons, and requires a force of 250 pounds to drag it with a uniform velocity, then the tractional resistance is $\frac{250}{18}$, or $13\frac{8}{9}$ pounds per ton.

Examples:

1. A train is made up of *12* cars each representing a total load of *23* tons. If the tractional resistance is *10* pounds per ton load, what force will be necessary to keep the train moving with a uniform velocity?

2. What horse-power will be required to keep this train moving at the rate of *30* miles an hour?

3. If this train comes to an *8%* grade what horse-power would be required to carry the train up the grade with a velocity of *4* miles an hour?

Consider that the friction is the same on the slope as on the level.

Anti-Friction Wheels and Ball Bearings. When a wheel supports a vehicle of any sort in such a manner as to require bearings, the resistance encountered in moving the vehicle includes not only the rolling friction at the circumference but also some sliding friction in the bearings. To get rid of this sliding friction, or rather to reduce it to a minimum, a number of devices, such as anti-friction wheels, roller bearings and ball bearings, are used under varying conditions.

The manner of using anti-friction wheels is shown in Fig. 128. Suppose that a weight W is carried on the circumference of a wheel C. The axle S of this wheel, instead of revolving in fixed bearings, is supported by the rims of two other wheels, C_1 and C_2, the axles of these, however, being in fixed bearings. As S revolves C_1 and C_2 also revolve, but with a much smaller angular velocity Now the total load and the friction in the bearings of C_1 and C_2 may even be greater than would have existed in the bearings of S if fixed bearings had been used at that point instead of the anti-friction wheels. But the rate at which the axles of C_1 and C_2 move in their bearings is so slow that very little

FRICTION.   177

work is done during each revolution of S, and hence it happens that the energy consumed by sliding friction in the bearings of C_1 and C_2 *plus* the rolling friction of S on the rims of C_1 and C_2 is less than what would have been used up by the sliding friction of S placed directly in fixed bearings.

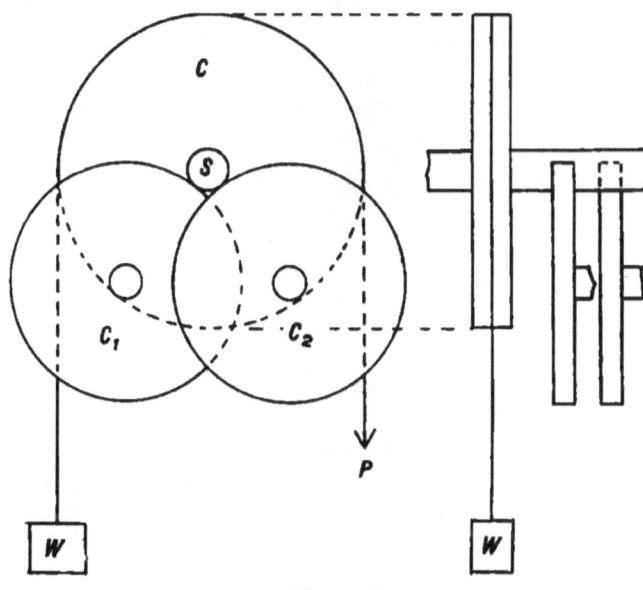

Fig. 128.

Ball Bearings have become so well known through their extensive application to bicycles that the illustration in Fig. 129 needs no special description. If the axle S revolves in the direction of the hands of a clock, the balls will all revolve in a counter-clockwise direction, and will also progress in a train as suggested by the large arrow A, rolling on the inner surface of the fixed bearing, represented by the portion B. From this it will be seen that where the balls touch S and B the friction is entirely rolling friction. But if we consider the points of contact between the balls

178   THEORETICAL MECHANICS.

themselves it will be observed that at all these points there is sliding friction. For example, in Fig. 130, which represents on a larger scale the adjacent balls b and b_1 of Fig. 129,

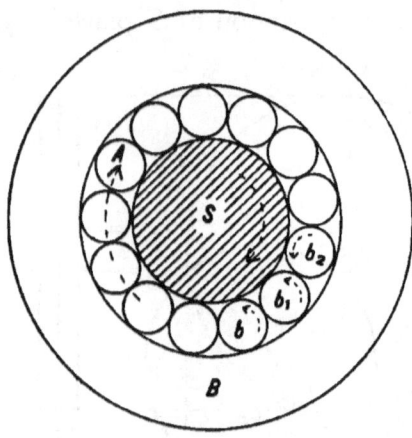

Fig. 129.

the point p of ball b is shown to be moving upward, and at the same instant the point p_1 of ball b_1 is necessarily moving downward, if the two balls are revolving in the same direction.

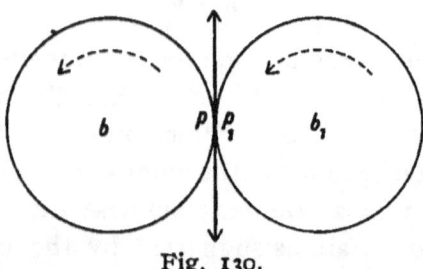

Fig. 130.

The occurrence of this sliding at each point of contact between the balls is sometimes urged as an argument against the use of ball bearings, notwithstanding that their great efficiency has been so widely demonstrated in practice. While it is true that the balls must slide past each other, it is also true that there is no great pressure between them for the reason that the weight on the axle S tends to spread the balls outward in the direction of the radial lines shown in Fig. 131, thus tending to separate them from each other in the manner illustrated by the small dotted circles in the

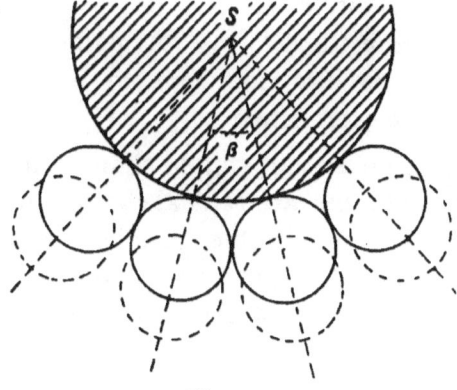

Fig. 131.

figure. Hence, in the absence of a very appreciable pressure at the points of contact, the friction between the balls as they slide past each other cannot be great.

This sliding friction, however, such as it is, exists at every point of contact, so that the greater the number of balls used the greater the total friction. Hence the advantage of using a small number of large balls, rather than a larger number of smaller ones. The larger balls also have the advantage that their tendency to be pressed away from each other is greater because of the greater angle β, indicated in Fig. 131.

Lubricated Surfaces Depart Widely from the Ordinary Laws of Sliding Friction. When two surfaces as in shaft bearings are separated by a lubricant, such as an oil, a grease, or plumbago, the simple laws of sliding friction are greatly modified, especially as regards kinetic friction, or friction of motion. The friction is no longer between the given solid surfaces entirely, or under some circumstances at all. Much depends (1) upon the lubricant itself,—the kind and quantity, and the manner in which it is applied ; and (2) upon the relative velocity of the two surfaces, and the pressure between them. (3) A temperature change at the two surfaces is also sufficient to change the entire relations otherwise established.

Notice that these are not simple changes in the values of constant coefficients, but are radical changes of fundamental laws, giving rise to variable coefficients for the same surfaces. At first thought one would say that if the lubricant adheres to each of the two rubbing surfaces, the motion must be a sliding of one part of the lubricant on another part, and hence it would be sufficient to substitute for the coefficient of friction between the given surfaces the coefficient of oil on oil, or plumbago on plumbago, etc. This idea, however, is not upheld by careful observations that have been made.

Scientific investigation has not furnished altogether conclusive results as to the laws that apply to the friction of lubricated surfaces. It is certain, however, that the friction under these circumstances is not a constant ratio of the pressure between the surfaces, as in the absence of lubrication. This *ratio* (the coefficient of friction) for lubricated surfaces decreases as the pressure increases and continues to do so up to a certain limit, beyond which it becomes greater again up to the time "cutting" or abrasion begins.

For unlubricated surfaces the difference between static and kinetic friction (friction to start the motion as contrasted with friction during the motion) is never very great. Between lubricated surfaces the friction at starting is much greater in proportion,—until the motion of the surfaces has carried the lubricant to the bearings, from which it has been squeezed out during rest. This is especially the case in the bearings of heavy machinery, or where the lubricant is unduly limpid.

The controlling conditions—kind of lubricant, its quantity and manner of application; temperature; velocity; and pressure in bearings—are inter-related in such a complex manner that it is practically impossible to express these relations in any but a very general way and with more than approximate correctness.

In general, the lubricant should adhere to the surfaces sufficiently to be constantly dragged in to the rubbing area as fast as needed. It can be forced in by external pressure, under some circumstances.

If the pressure between the bearings is very great the the lubricant is forced out, especially if it is too thin, and in such cases it is better to use grease or plumbago, or "heavy" oil. In small bearings, under light pressure, a thick, viscous oil would be needlessly cohesive.

Likewise, at varying velocities the amount of lubricant drawn into the rubbing area by the motion of the bearings, and the general effect, is very variable. At moderate speeds the coefficient of friction decreases as the velocity increases, but for very high speeds the opposite is true. And furthermore, the speed at which this change from a decreasing to an increasing coefficient takes place is different for high and low pressures, and for high and low temperatures and for different lubricants.

Conditions that are satisfactory at one temperature might be actually reversed at even a slightly different temperature. The properties of oils change greatly under the influence of heat, and no two oils change in the same manner or degree. In general, the lighter oils used for high speeds and low pressures give better service at comparatively high temperatures, while the more viscous oils and greases used for high pressures lose their desired viscosity when heated. If the temperature becomes very high the organic (animal and vegetable) oils may be decomposed, liberating injurious acid compounds that attack the material of the bearings. In steam chests and cylinders only or mainly mineral oils, which do not suffer decomposition under heat, should be used.

On the whole the friction of lubricated surfaces presents many questions for scientific investigation, and it is not safe to draw very broad conclusions from the superficial and limited observations that are likely to be met with in this connection in ordinary practice.

Friction of Ropes, Belts and Cables. When a flexible belt or cord is wrapped upon a cylindrical surface, any tension exerted on the belt or cord produces a normal pressure against the cylindrical surface at every point in the arc of contact. This obviously will cause friction, and while this friction will be in accordance with the general laws already deduced, it will also involve conditions and circumstances requiring special consideration.

The hitherto simple law of sliding friction ($F = \phi P$), when extended to cover the case of belts and ropes on pulleys and sheaves, becomes so complex that it cannot be expressed without the use of exponential or logarithmic formulæ.

FRICTION. 183

Suppose that the circle C in Fig. 132, represents a cylinder screwed firmly to a horizontal board so that it *cannot rotate*.* A cord is placed around the cylinder in the manner shown, affording a contact for a half circumference, or 180°. One end leads to a spring balance fastened to the board. Now if a tension T be cautiously exerted on the free end of the cord (avoiding sudden jerks) it will be found by reading the spring balance that the tension t in the other

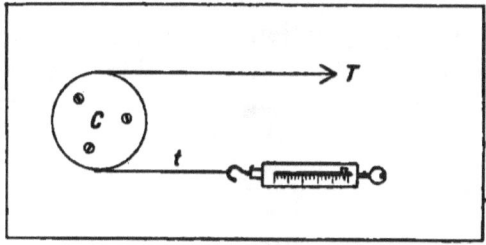

Fig. 132.

part of the rope is much less than T. If the tension T be increased, t will also be found to be greater in the same proportion. That is, if the two ends of the cord are kept parallel (so as not to change the length of the arc of contact), the ratio $\dfrac{t}{T}$ will remain constant; t will always be a definite portion of the applied tension T. Small t is less than T because of the friction of the rope or belt on the cylinder, and this friction is obviously equal to the difference

---

*The transmission of power by means of belts and ropes presupposes that the pulley moves with the belt—a sort of rolling contact. Some fundamental conceptions, however, can be grasped more readily by first regarding the pulley as immovable. In fact the maximum power transmissible is determined by the conditions at the time of slipping. If the pulley moves with the belt the coefficient is one of static friction; if the belt is intended to slip on the pulley, as in the friction brake, the coefficient is for kinetic friction. The relations above deduced are the same in both cases, the only difference being in the magnitude of the constant coefficients for static and kinetic conditions.

$T - t$. For example, if $T = 5$ pounds, and $t = 1$ pound, the friction of the cord on the cylinder is $F = 4$ pounds. If we increase T to 15 pounds, then t will be 3 pounds and $F = T - t = 12$ pounds. Expressed in different form, if

$$\frac{t}{T} = \frac{1}{5}, \text{ then } \frac{F}{T} = \frac{4}{5}, \text{ or } F = \frac{4}{5} T.$$

Now it is this fact—that the friction is itself a function of the force that overcomes it—which gives rise to the complex mathematical relations that have to be dealt with in studying the friction of belts and ropes. In the simple phenomenon of sliding friction, the force which overcomes the friction between the weight W (Fig. 133) and

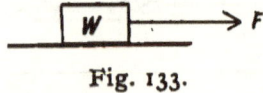

Fig. 133.

the surface upon which it rests has no effect to change the magnitude of this friction one way or the other. In the case of the friction of the rope on the cylinder, on the contrary, the very tension that is intended to make the cord move is what causes the friction on the cylinder, and the greater the tension the greater the friction. Hence, in practice, the more power to be transmitted the tighter the belt, and the wider and thicker it must be, to stand the necessary tension.

The Friction of a Belt Depends upon the Arc of Contact. The friction of a belt on a pulley is not only a function of the applied tension, but depends also upon the magnitude of the arc of contact. In Fig. 132 the rope or belt was in contact with one-half the circumference of the pulley. If now the end to which the tension is applied has a direction such that the rope is in contact with an arc β, less than the semi-circumference, then the ratio $\frac{t}{T}$ is no longer $\frac{1}{5}$, as in the last paragraph, nor is the friction $F = \frac{4}{5} T$. The effect of reducing the arc of contact will be to diminish the friction, leaving a greater proportion of the applied tension, T, to be propagated along the cord or belt to the spring balance, to cause tension t.

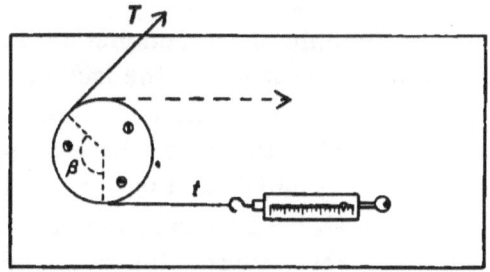

Fig. 134.

Hence, the friction depends upon the extent to which the belt envelops the pulley. This is not, as it might appear, in violation of the law asserting that the sliding friction between two objects is independent of the area of contact. The friction between two surfaces is independent of the area of contact, *provided that in changing the area we do not at the same time change the pressure between the surfaces.* For instance, if a brick is turned from a flat side to an edge, the area of the surface of contact is changed, but the total normal pressure is still the weight of the brick; the area is less, but the pressure per unit area is greater. But if several bricks are fastened together in a "train" as in Fig. 135, the pressure per unit area is unchanged, and hence the increased area of contact is accompanied by an

186   THEORETICAL MECHANICS.

increase of total pressure, thereby multiplying the total friction in the same ratio. The friction of a belt on a pulley is analogous to the train of bricks. Every degree added to the arc of contact adds to the total pressure of the belt against the pulley and thus increases the friction.

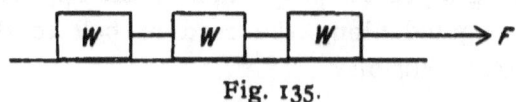

Fig. 135.

Friction of a Flexible Cord Completely Enveloping a Cylindrical Surface. The laws for the friction of belts and ropes apply not only to ordinary transmission of power from pulley to pulley, but also to cases where a rope is coiled several times around a cylinder, as the rope on the drum of a hoisting engine or a ship's hawser around a post. If there is but one turn of the rope around the cylinder, the arc of contact is 360°, for two turns 720°, etc. The tension that must be exerted at the free end to prevent slipping is, as before, the difference between the applied tension T and the friction. With this same applied tension, if we take a second turn of the rope around the cylinder, the value t left after one turn may be regarded as the "applied tension" for the next turn, so that if t_2 is the tension that remains in the free end after two turns, then

$$\frac{t_2}{t} = \frac{t}{T}, \text{ or } \frac{t_2}{T} = \left(\frac{t}{T}\right)^2.$$

If there were three turns of the rope the tension in the free end, t_3, would be found from the relation,

$$\frac{t_3}{t_2} = \frac{t_2}{t} = \frac{t}{T}, \text{ or } \frac{t_3}{T} = \left(\frac{t}{T}\right)^3.$$

For example, fasten a cylinder of 3-inch or 4-inch diameter to a board, and screw this board to the wall. A stout cord is substituted for a rope. The upper end of the cord is hooked onto a sensitive spring balance, and the applied tension is produced by weights. The ratio $\dfrac{t}{T}$ for one turn is nearly constant for all tensions, but not absolutely so, because on account of the twisting of the strands the flexibility of the rope is different for different values of T. Suppose that, roughly, $\dfrac{t}{T} = \dfrac{1}{6}$. Now if we take a second turn around the post this value, t or $\dfrac{T}{6}$, must be taken as the applied tension for the second turn, and the tension on the balance will now be (approximately) $\dfrac{1}{6} t$ or $\dfrac{1}{36} T$. For three turns the balance will read

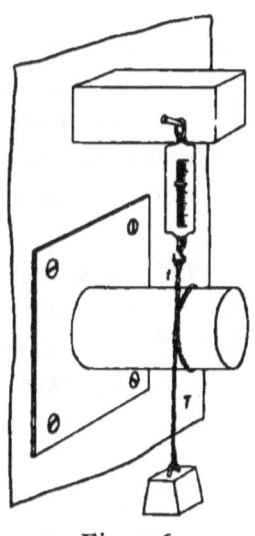

Fig. 136.

$$\dfrac{1}{6} \times \dfrac{1}{6} \times \dfrac{1}{6} T, \text{ or } \dfrac{1}{216} T.$$

In general terms, if t_n is the tension left after n turns of the rope, then

$$t_n \propto \dfrac{T}{K^n},$$

where the exponent n is the number of turns, and K the constant depending on the nature of the surfaces.

This is sufficient to show in a general way that the various factors to be considered in the friction of belts and ropes are related to each other in such a way as to give rise to mathematical expressions of an exponential character. In the next paragraph these relations are expressed by means of logarithms in a usable form for the ordinary practice of belting.

Mathematical Expression of the Friction of a Belt or Rope in Terms of the Tensions and the Arc of Contact.

Having shown in a general way that the friction of a belt or rope on its pulley depends upon (1) the tension of the belt, and (2) the magnitude of the arc of contact, it may be well to express these relations in exact mathematical form. We can only outline the method of deducing these expressions.*

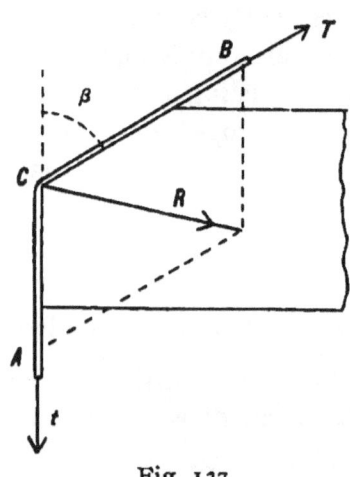

Fig. 137.

Suppose that ABC (Fig. 137) represents a belt bent at an angle β over the edge of a block of iron, the angle β being determined by its supplementary angle BCA in the block. The applied tension T is sufficient to overcome the friction necessary to slide the belt over the angle at C and to leave a tension t in the other end of the rope, or $T = F + t$, where F represents the friction at C. The friction at C is caused, of course, by the normal pressure of the belt against the block at that point, and this normal pressure is merely the resultant, R, of T and t. There is no friction on the *faces* of the block in contact with the belt, because there is no pressure of the belt against those faces. Hence we have $F = \phi R$. But

$$R = \sqrt{T^2 + t^2 - 2Tt \cos \beta} \,; \qquad **$$

whence

$$F = \phi \sqrt{T^2 + t^2 - 2Tt \cos \beta}.$$

---

*For full demonstration see Cromwell's "Belts and Pulleys", Section X,—Wiley & Sons, New York, Publishers.

**See Kinematics, p. 14. Notice the negative sign of the third term under the radical sign,—due to the fact that BAC is $180° - \beta$.

FRICTION.   189

Using this equation and the relation $F = T - t$, by proper algebraic and trigonometric transformations, we may deduce the expression

$$\frac{T}{t} = 1 + 2\phi \sin \frac{\beta}{2}.$$

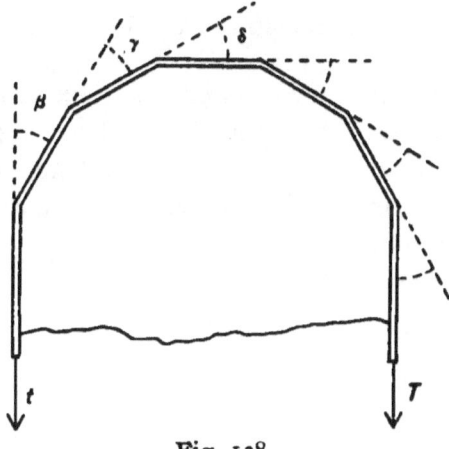

Fig. 138.

Now what is true of this β will be true in a general way for each of the angles, β, γ, δ, etc. in Fig. 138.

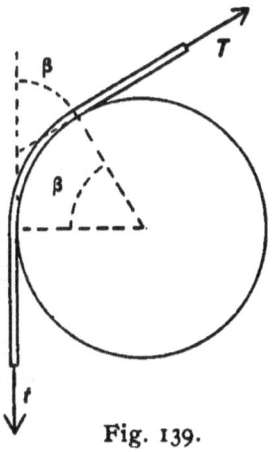

Fig. 139.

From a large number of finite angles we can reason to an infinite number of infinitesimal angles, such as would constitute the total angle β in Fig. 139, illustrating a belt and pulley. By this process we would arrive at the following general formula, representing the relations between the tensions T and t, and the angle β, and expressed as a common logarithm:

$$\log \frac{T}{t} = 0.007578\, \phi\, \beta,$$

in which

β is expressed in degrees;
ϕ is the coefficient of friction;
T is the tension of the tight side; and
t is the tension of the loose side.

From this formula we can deduce an expression showing directly the relation between the friction, F; the applied tension, T; the arc of contact, β; and the coefficient of friction Expressing the above formula as an anti-logarithm, we get the expression

$$\frac{T}{t} = \log^{-1} 0.007578\, \phi\, \beta\,;$$

or, inverting,

$$\frac{t}{T} = \frac{1}{\log^{-1} 0.007578\, \phi\, \beta}.$$

Now remembering that t is that part of T which is left over after overcoming the friction, and putting the expression $F = T - t$ in the form

$$F = T\left(1 - \frac{t}{T}\right),$$

and substituting for $\frac{t}{T}$ the value deduced above, we get

$$F = T\left(1 - \frac{1}{\log^{-1} 0.007578\, \phi\, \beta}\right). \qquad (11)$$

To use this formula in practice, it is necessary to know the coefficient of friction for the various kinds of ropes, belts and pulleys in common use.

FRICTION.   191

For leather belts on iron pulleys, $\phi = 0.40$

For leather belts on leather-covered pulleys, $\phi = 0.45$.

For hemp rope on iron in semi-circular groove (Fig. 140), $\phi = 0.30$. (Used for small powers only).

For hemp rope on iron in V-groove, with straight sides inclined at angle of 45° (Fig. 141), $\phi = 0.70$.

For steel cable in groove lined with leather (Fig. 142), $\phi = 0.24$.

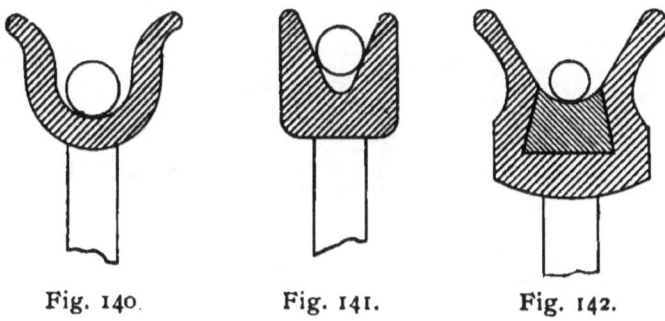

Fig. 140.   Fig. 141.   Fig. 142.

Examples:

1. *A leather belt runs over a cast iron pulley enveloping an arc of 140°. What is the friction between the belt and the pulley if slipping begins when the tension in the driving side of the belt is 90 pounds?*

2. *A rope belt running over a pulley in a V-groove has an arc of contact of 180°. A force of 200 pounds applied to the perimeter of the wheel is just sufficient to prevent it from rotating so that the rope has to slip over the pulley. What is the value of T?*

3. *A leather belt envelops an arc of 150° on a leather-covered pulley. What will be the friction when T = 200 pounds?*

4. *The friction of a leather belt enveloping an arc of 190° on a leather-covered pulley is 115 pounds. What is the tension in the driving side of the belt, and what is the tension in the slack slide?*

Measurement of Power Transmitted by Belts.

In the ordinary continuous transmission of power by means of an endless belt, the angle of contact on each of the two pulleys depends upon their relative diameters and their distance apart. The maximum power that can be transmitted from one pulley to the other through the belt depends upon the friction at that pulley where slipping will first occur. If both pulleys are of the same material this will be at the pulley having the smaller arc of contact.

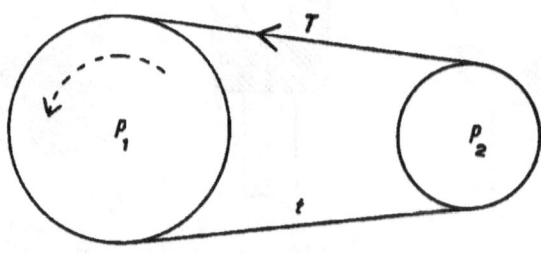

Fig. 143.

In Fig. 143, illustrating pulleys of 20-inch and 10-inch diameters respectively, if p_1 is the driving pulley, it will not be possible for the belt to transmit to the second pulley, p_2, all the power that the driving pulley would be capable of giving to the belt. The friction of the belt on p_2, in causing the motion of the pulley, is equivalent to a force of the same magnitude acting at any single point on the perimeter of p_2 in a tangent direction. And it should be remembered, again, that this maximum force is not T, the tension in the belt, but only a part of it,—that is, the friction F, or

$$T\left(1 - \frac{1}{\log{}^{10} 0.007578 \, \phi \, \beta}\right).$$

FRICTION.

If the diameter of p_2 is d inches, then for each revolution of p_2 the maximum work that can be done, or the maximum energy that can be taken from the belt by this pulley, is $F \times \dfrac{\pi d}{12}$ foot-pounds (provided F is in pounds).

If this pulley has n revolutions per minute, the maximum H. P. that can be transmitted to it from the belt is

$$F \times \frac{\pi d}{12} \times \frac{n}{33,000}.$$

Examples:

1. In the case of the pulleys and endless belt just referred to, if p_1 and p_2 have diameters of 20 inches and 10 inches respectively, and are 12 feet apart between centers, what is the magnitude of the arc of contact of the belt on each of the pulleys?

2. If the shaft of p_1 revolves 165 times per minute, what is the maximum power that can be conveyed to the shaft of p_2, if the tension of the driving side of the belt is 211 pounds when slipping commences?

Methods of Increasing the Efficiency of Belts. Anything that will increase the tension of the belt, or the arc of contact, or the coefficient of friction, will increase the maximum power transmissible by the belt.

We have already seen that the coefficient is increased from 0.4 to 0.45 by covering the pulley with leather. The two pulleys should not be near enough together to interfere with the flexible working of the belt. If the pulleys are of unequal size the arc of contact on the smaller will be greater, the farther apart the pulleys. If the slack side of the belt is uppermost, the sag of the belt will increase the arcs of contact.

By the use of a so-called "idle" or tightening pulley, as illustrated in Figs. 144 and 145, it is possible to increase

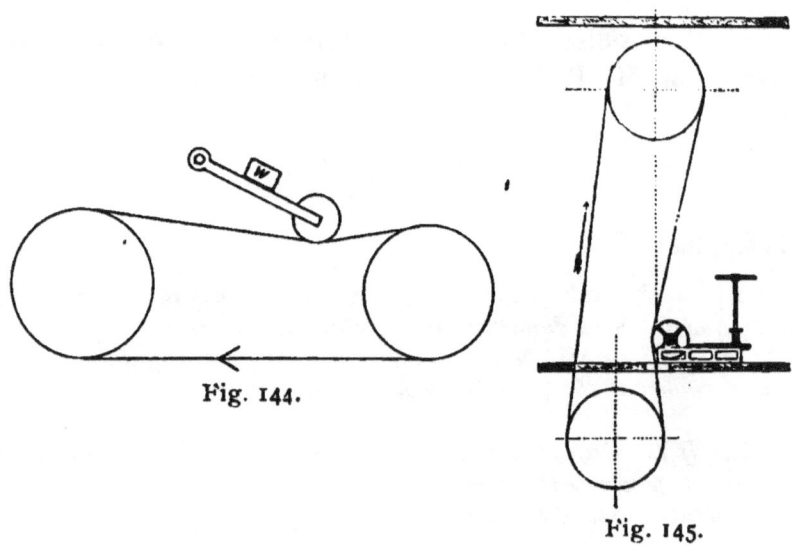

Fig. 144.

Fig. 145.

both the tension and the arcs of contact at the same time. Fig. 144 illustrates the method of using a tightening pulley on horizontal belts; Fig. 145 is for vertical belts.

Favorable results have been reported of a simple and apparently practical method of increasing the normal pressure between the belt and the pulley without subjecting the belt to excessive tension. A strip of canvas (or other suitable material) narrower than the belt, together with a short length of stout sheet rubber to be inserted between the ends of the canvas strip, is made into a band. This band is stretched garter-like over the belt.

With a little thought it will be seen that the action of any belt necessitates some slipping. One side of the belt being tighter than the other, as each minute portion of the belt progresses along the perimeter of the driven pulley to a position of greater tension, it is gradually stretched out, and hence must slip a little. As it again approaches the slack side of the other pulley—the driving pulley—it contracts and slips in the opposite direction. The elasticity of the material, therefore, is a factor to be considered in the easy working of the belt. This constant slipping generates heat, causing a gradual burning of the belt.

It will also be observed from formula 11, p. 190, that the friction depends on T, ϕ and β, and not upon the width of the belt or diameter of the pulley. This applies down to the limit at which the lack of perfect flexibility in the belt would prevent it from conforming properly to the surface of too small a pulley.

www.ingramcontent.com/pod-product-compliance
Lightning Source LLC
Chambersburg PA
CBHW020919230426
43666CB00008B/1498